Lab-on-a-Chip Devices and Micro-Total Analysis Systems

Jaime Castillo-León • Winnie E. Svendsen
Editors

Lab-on-a-Chip Devices and Micro-Total Analysis Systems

A Practical Guide

Springer

Editors
Jaime Castillo-León
Sol Voltaics AB
Lund
Sweden

Winnie E. Svendsen
Department of Micro- and Nanotechnology
Technical University of Denmark
Kgs. Lyngby
Denmark

ISBN 978-3-319-08686-6 ISBN 978-3-319-08687-3 (eBook)
DOI 10.1007/978-3-319-08687-3
Springer Cham Heidelberg New York Dordrecht London

Library of Congress Control Number: 2014953388

Printed on acid-free paper

Springer is part of Springer Science+Business Media (www.springer.com)

Preface

Lab-on-a-Chip Devices and Micro-Total Analysis Systems: A Practical Guide is a result of the research and teaching activities of a group of scientists worldwide who wrote the chapters of this book dealing with the design, simulation fabrication, and application of lab-on-a-chip and micro-total analysis systems. In a very didactic way this book covers the most important steps for the development of lab-on-a-chip devices and presents the state of the art in the use of new materials and novel uses of this technology.

We would like to thank all the authors contributing to this book for their effort in helping us to put together this work. We also would like to thank Springer for inviting us to publish this book and all the support during its preparation.

We hope that this work will motivate a broad audience of students and researchers interested in lab-on-a-chip and micro-total analysis systems so as to consider these technologies as an alternative in their research. Here they will find the advantages and challenges when working with these devices as well as inspiring examples of work carried out using microfluidic platforms.

Lund, Sweden — Jaime Castillo-León
Kongens Lyngby, Denmark — Winnie E. Svendsen

Contents

Chapter 1
Microfluidics and Lab-on-a-Chip Devices: History and Challenges

Jaime Castillo-León

Abstract The rapid advances in microfabrication and nanofabrication in combination with the synthesis and discovery of new materials have propelled the drive to develop new technological devices such as smartphones, personal and tablet computers. These devices have changed the way humankind interacts and communicates and this change occurred very quickly due in part to decreased production and commercialization costs. As a result, not only nations with powerful economies but also emerging economies and poor countries can get access to these technologies and experience new ways to interact and instantly communicate what is happening around us. Following the advances of all these communication devices as well as those in microfabrication and nanofabrication and the emergence of new materials, technologies such as lab-on-a-chip (LOC) and micro total analysis systems (microTAS) were also boosted, albeit at a slower pace. LOC and microTAS applications have principally been utilized in the biomedical, food and environmental fields. But lately they have also found their place in the synthesis of new chemical compounds and the fabrication of nanostructures. It has become obvious that the LOC and microTAS technologies need to join forces with those behind the new communication devices which provide sources of power, detection and data transmission complementing the features that lab-on-a-chip and microTAS platforms can offer. An increasing number of microfluidic-based devices, developed both in small start-ups and large pharmaceutical and biomedical companies, is being released and entering the market. This chapter offers an overview of the first events in the history of LOC and microTAS devices, the biggest achievements and the challenges that still need to be overcome in order to accelerate the use of this technology.

J. Castillo-León (✉)
DTU Nanotech, Technical University of Denmark, Kongens Lyngby 2800, Denmark

Sol Voltaics AB, Scheelev. 22, Lund 223 63, Sweden
e-mail: jcl@solvoltaics.com

J. Castillo-León, W.E. Svendsen (eds.), *Lab-on-a-Chip Devices and Micro-Total Analysis Systems*, DOI 10.1007/978-3-319-08687-3_1

1.1 History

The origin of microfluidics devices fabricated using the technology of micromechanics is usually associated with the late 1960s and the works on gas chromatography at Stanford University and the ink jet printer nozzles developed at IBM. But hundreds of years earlier, researchers were already curious about the information they could extract from body fluids and were trying to understand the behavior of liquids confined in small diameters. Hippocrates (400 BC), Galen (200 AD), and Theophilus (700 AD) among others were interested in analyzing urine samples as a means of interpreting the functions of the human body (Gazzaniga 1999; Kouba et al. 2007; Wallis 2000). They performed these analyses bedside and tried to match the color and appearance of urine with symptoms, mainly fever and psychiatric disturbances. They did not have the possibility to use microfluidic chips; instead, they investigated the urine samples in examination flasks known as "matulas" verifying the color and appearance of the sample (Wittern-Sterzel 2000). They even tried to standardize this analysis method by drawing diagrams and pictures of "matulas" of colored urine to be used as a diagnostic tool (Kouba et al. 2007).

Hundreds of years later, James Jurin, a British physician and natural philosopher (1684–1749), Jean Léonard Marie Poiseuille, a French physicist and physiologist, and Gotthilf Heinrich Ludwing Hagen (1791–1884), a German physicist and hydraulic engineer, worked on describing the behavior of fluids that were confined in glass containers of small diameters. They published their studies on capillarity, in the case of Jurin, and formulated the famous Hagen–Poiseuille equation that describes the pressure drop in a fluid flowing through a cylindrical pipe establishing the basis of the theory explaining the behavior of fluids at reduced scales.

Today, thanks to the advances in microfabrication and nanofabrication and the synthesis of new materials, researchers can confine fluids into microchannels or nanochannels instead of "matulas" and have theoretical tools and simulation software that help them predict and understand the behavior of fluids contained in microfluidic devices. Currently, not only body fluids but also cells, tissues and even a whole organ can be investigated inside microfluidic chips.

As mentioned above, thanks to the new fabrication methods and available tools, the end of the 1960s saw the development of a first prototype of a quadrupole GC/MS by Finnigan Instrument Corporation who delivered it to Stanford and Purdue University. This first instrument cost 100,000 USD and it was controlled by a minicomputer (Brock 2011). A couple of years later, following the idea of William Thomson (later knows as Lord Kelvin), who in 1867 patented the use of electrostatic forces to control the release of ink drops onto paper, in 1951, Siemens produced the first continuous inkjet printer used in medical strip chart recorders. Then, in 1976, researchers at IBM produced the first commercially available continuous forms of laser printers (Bassous et al. 1977).

In 1993, Andreas Manz published a fundamental article describing the fabrication of a miniaturized capillary electrophoresis-based chemical analysis system on a chip (Harrison et al. 1993). In this work, a microchip of 1 cm by 2 cm was micromachined in glass and electroosmotic pumping was employed to drive the fluid flow and perform electrophoretic separation on-chip.

Just 1 year after the publication of Manz's work, the first International Conference on Micro-Total Analysis Systems took place in Enschede, The Netherlands, and has since then strengthened its position as the premier forum for reporting the latest research results in microfluidics, Lab-on-a-Chip, microfabrication, nanotechnology, integration and detection technologies for life science and chemistry (www.cbmsociety.org).

In 1998, Professor George Whitesides, one of the leaders and more active members of the microfluidics and Lab-on-a-Chip community, published an article on the rapid prototyping of microfluidic systems in polydimethylsiloxane (PDMS). This publication described the design and fabrication of microfluidic chips in a transparent elastomeric material in less than a day using a photolithography technique. It had a large impact on the microfluidics community and pushed the field forward providing researchers with an easy and fast fabrication method that has been widely used in various studies (Duffy et al. 1998; Xia and Whitesides 1998).

Two years later, the group of Dr. Richard D. Fair at Duke University (USA), published an article on the rapid actuation of discrete liquid droplets accomplished by direct electrical control of the surface tension (Pollack et al. 2000). This is one of the first publications describing a liquid handling technology known as digital microfluidics. This handling technique allows a controlled manipulation of picoliter- to microliter-sized droplets independently addressed on an open array of electrodes. This new method had a huge impact on chemical and enzymatic reactions, DNA-based applications, immunoassays, clinical diagnostics, cell-based applications, and proteomics due to its flexible device geometry, simple instrumentation, and easy integration with other technologies (Choi et al. 2012).

Seven years later, a new fabrication method was published by Prof. Whiteside's group describing the use of paper as a new, portable, low-cost alternative for the development of bioassay microfluidic devices (Martinez et al. 2007). By patterning paper, Whitesides and coworkers created well-defined millimeter-sized channels and hydrophilic and hydrophobic zones enabling the transport of fluid to desired locations without the need for pumps due to the capillarity effect. This new fabrication method offered a cheap alternative for the development of biosensing platforms at a very low cost that could be applied in countries with limited resources (Yetisen et al. 2013). More details regarding the use of paper-based microfluidic devices can be found in Chap. 7.

A couple of years ago, a technology called 3D imprinting, developed in the early 1980s at Ultra Violet Products in California by Charles Hull, probed the possibility of fabricating miniaturized fluidic reactors for the chemical syntheses of secondary amines, gold nanoparticles, and large polyoxometalate clusters (Kitson et al. 2012). One of the more attractive features of this new technology is the possibility to fabricate devices with three-dimensional features in a quite simple way which is expected to help the manufacture of large-scale integrated multilayer devices and their standardization between laboratories. This could possibly start the third decade of microfluidic revolution (Gross et al. 2014; Lee 2013).

A clear example of the impact of the 3D imprinting technology can already be found in studies involving cells and organs-on-chips (Kim et al. 2008; Tasoglu et al. 2013; Young 2013). By using 3D printing, microfluidic environments mimicking the in vivo situation can be fabricated providing a "closer to reality" environment that is being used in pharmaceutical labs to evaluate new drug candidates. Recently, the idea of building a body-on-chip device has been presented (Moraes et al. 2013; van der Meer and van den Berg 2012). The idea involved connecting compartmentalized environments in which different cells are cultured to mimic multiple organs that will be interconnected or isolated depending of the application in order to recreate a human body. This will have a big impact on the biomedical field, offering researchers a new test tool where new medicines and treatments can be tested.

As will be shown throughout this book, microfluidics and lab-on-a-chip devices offer a long list of attractive advantages for research in numerous fields requiring a low consumption of sample and reagents, portability, low-energy consumption, rapid response and multiplexing analysis, etc. However, challenges regarding its application at a large scale, the possibility of work with bigger volumes, the fabrication of low-cost and user-friendly devices and its integration with other analytical techniques are issues that need to be solved in order to accelerate the irruption of these technologies in our daily lives in a manner similar to that of smartphones and personal computers. In fact, there are many recent reports of LOC devices integrated in smartphones. The idea is to connect the microfluidic devices with the smartphone and take advantage of the features available in the latter, i.e., the power source, data transmission, and optical detection system (Mark et al. 2012; Yetisen et al. 2014).

As will be seen in the chapters ahead, several strategies to overcome these challenges are being successfully implemented and resulting in the commercialization and fabrication of new microfluidic, lab-on-a chip and microTAS devices (Araci and Brisk 2014; Chin et al. 2012) as presented in Table 1.1.

Researchers that are interested in learning, discussing, and sharing experiences regarding the use of microfluidics have multiple options for this through journals, conferences, blogs, and forums focusing on microfluidics, some of which are presented in Tables 1.2 and 1.3. These alternatives offer a great environment to exchange ideas, tips, and experiences and hopefully create a community for starting new collaborations between scientists from different backgrounds (chemistry, physics, biology, material science). Such collaborations are one of the reasons that microfluidics has been able to expand its impact and should definitively increase this impact on our lives.

Table 1.1 Companies working on the design, fabrication and/or application of microfluidic, lab-on-a-chip and μTAS devices

Company	Product	Description	Link
abbott	i-STAT portable handheld	Detection of cardiac markers, blood gases, electrolytes, lactate, and clotting time. Wireless data transmission, requires 2–3 drops of blood to perform a test	http://www.abbottpointofcare.com/Products-and-Services/iSTAT-Handheld.aspx
Advanced Liquid Logic	LSD-100	Bioassay automation platform capable of simultaneously performing 4 enzymatic activity assays on 40 dried blood spot, use a digital microfluidic cartridge	http://www.liquid-logic.com/lsd-100
Agilent Technologies	2100 Bioanalyzer	Provides sizing, quantitation, and quality control of DNA, RNA, proteins, and cells on a single platform. Requires 1–4 mL	http://www.genomics.agilent.com/en/Bioanalyzer-System/2100-Bioanalyzer-Instruments/?cid=AG-PT-106&tabId=AG-PR-1001
Alere	Afinion™ test	Microfluidic cartridge for measurement of glycated hemoglobin, requires 1.5 μL sample	www.alere.co.uk
	epoc®	Microfluidic card for the measurement of electrolytes and blood gas	
Alexeter Technol. LLC	RAID	Multi-strip assays for anthrax, ricin, bot tox, SEB, plague, tularemia, brucella, and pox	http://www.alexeter.com/biow/index.asp
Aline Inc.	Microfluidic slides, cuvettes	Design, development and manufacture of lab-on-a-chip and microfluidic products	www.alineinc.com
ANSYS	ANSYS CFD	Computational fluid dynamics software	www.ansys.com
Avalance Bioetch AB	Multifunctional pipette	Pneumatic control unit for localized delivered of small volumes to cells	www.avalancebiotech.com
Bartels Mikrotechnik GmbH	Microfluidic components: micropumps, check valves, tubing	Piezo actuated membrane micro pumps, fabricated in polymers: polyphenylsulfone, polypropylene	www.micro-components.com

(continued)

Table 1.1 (continued)

Company	Product	Description	Link
bi.Flow Systems GmbH	Flex,flow microscope slides	Microscope slides providing 6 liquid reservoirs with integrated micropumps which can be individually emptied in a controlled way	www.biflow-systems.com
Biomerieux-diagnostics	bioNexia rapid tests	Point-of-care rapid tests, take less than 10 min to perform. Used for diagnostics of HIV, malaria, hepatitis B, influenza, screening of bladder cancer, human cardiac Troponin I, C-reactive protein, human chorionic gonadotrophin	http://www.biomerieux.com/en/innovation/point-care#bioNexia® rapid tests
Biophysical tools	Microperfusion system MPS8	Microperfusion system for quantitative biomedical research and high throughput	www.biophysical-tools.de
Cellix	Kima	iPod touch microfluidic pump, 300 μL dead volume, flow rate: 15–35 mL/h	http://www.cellixltd.com/index.php/products/microfluidic-pumps/kima
	Vena8 Fluoro+	Biochip for rolling and adhesion assays on protein coatings, 8 channels/biochip, 0.8 μL volume of each channel	http://www.cellixltd.com/index.php/products/biochips/vena8-fluoro
Comsol	Comsol Multiphysics®—Microfluidics module	Tools for studying and simulate microfluidic devices including lab-on-a-chip, digital microfluidics, electrokinetic and magnetokinetic devices and inkjets.	www.comsol.com
CorSolutions	Microfluidic handling components: pumps, probes, stations	Simulation software for the predictive 3D modelling and simulation solutions for microfabrication and nanofabrication technologies	www.mycorsolutions.com
Coventor	MEMS+	Simulation software	www.coventor.com
Daktari Diagnostics	Daktari CD4	Portable electrochemical-base blood taste	www.daktaridx.com

(continued)

Table 1.1 (continued)

Company	Product	Description	Link
Diagnostics For All (DFA) (nonprofit organization)	Paper-based microfluidic platforms	Paper-based diagnostic portable platforms with results available in minutes. Used for detection of bovine heat, milk spoilage, aflatoxine, DNA and RNA sequences among others	www.dfa.org
Dolomite	Liquid handling micro components: chips, pumps, valves, connectors, sensors, interfaces	Used to build both simple and complex systems	www.dolomite-microfluidics.com
DST diagnostics	FastCheckPOC®	Microfluidic test-cassette for the qualitative identification of allergen-specific IgE; requires 15 μL samples, results after 30 min	www.dst-diagnostic.com
Elveflow	Microfluidic flow sensors	Monitor flow down to 1.2 nL/s with a resolution of 1.5 pL/s	www.elveflow.com
ESI	CFD-ACE+	Simulation software; it enables coupled simulations of fluid, thermal, chemical, biological, electrical, and mechanical phenomena	www.esi-cfd.com
	Open FOAM	Open source CFD toolbox	www.openfoam.com
Espira Inc.	Plastic microfluidic disk	Automated pathogen detection unit to purify DNA or RNA within 20 min	www.espirainc.com
FlowJEM	Plastic microfluidic components	Thermoplastic microfluidic devices fabricated in polycarbonate, PDMS, cyclic olefin and acrylic	www.flowjem.com
Fluidigm	Plastic microfluidic chips and kits	Nanoliter scale plastic chips for single-cell gene expression and single-cell mRNA sequencing workflow	www.fluidigm.com

(continued)

Table 1.1 (continued)

Company	Product	Description	Link
Fluigent	Fluidic handling components: microfluidic flow control systems, software		www.fluigent.com
Focus Diagnostics	3M™ Integrated Cycler	Real-time PCR instrumentation, can process up to 96 patient samples per run in less than 80 min. Lab on a disk type of device	http://www.focusdx.com/product-catalog/3m-integrated-cycler
GenMark Diagnostics	e-Sensor® cartridge	Plastic microfluidic cartridge for the mixing of target DNA with the signal probe solution for electrochemical detection	www.genmarkdx.com
GeneFluidics	Asklepios system	Disposable microfluidic cartridge for multiplex electrochemical detection, requires less than 500 μL sample volume	www.genefluidics.com
Hμrel Corp.	Hμrelflow™	Microfluidic cell-based assay platforms for organ-on-chip studies	http://hurelcorp.com/products-services/microfluidics
Klearia	LabInGlass® microreactors and sensors	Microfluidic chips fabricated in glass	www.klearia.com
LeukoDx	Accellix platform	Cartridge-based system to provide CD4/HIV monitoring and sepsis	www.leukodx.com
Life Technologies	TaqMan®	Real-time PCR assay, uses 384-well microfluidic cards, 1 μL reactions	www.lifetechnologies.com
LTF Technologies	Microfluidic and lab-on-a-chip components: microreactors, pumps, valves, starter set	Microreactors for the production of chemicals on-chip	www.ltf-gmbh.com
Medimate	Medimate Minilab	Lab-on-a-chip device for lithium and sodium detection in blood	www.medimate.com
microfluidic ChipShop	Fluidic platforms and interfaces, diagnostic platforms	Offers both catalogue products and costume-specific product developments for the developments of LOC and μTAS devices	http://www.microfluidic-chipshop.com

(continued)

Table 1.1 (continued)

Company	Product	Description	Link
Micronics (Sony Group Company)	PanNAT system	PCR system for diagnosis of infectious diseases, 100 μL volume sample, results for up to 3 pathogens in less than 1 h	http://www.micronics.net/products/diagnostic-products/PanNAT
	ABORhCard®	Provides ABO and Rh blood type from a finger stick blood sample in 2 min	http://www.micronics.net/products/diagnostic-products/immunohematology
	H-filter® Access™ card	Used for sample processing for analytical chemistry, rare reagent isolation and recovery	http://www.micronics.net/products/research-and-development
	H-filter® and T-sensor® Access™ card	Used to demonstrate the quantitative detection of an analyte in a sample using a control solution	
Micronit Microfluidics	Fluidic chips, microreactors, chip holders, microfluidic starter kit	Custom microfluidic products in glass and polymers	www.micronit.com
Nanomix	Omega 3 platform	Disposable multiplex-assay cartridge containing immunoassay-enzymatic test for critical cardiac diagnosis	www.nano.com
Numeca Int.	Fine™	CFD and multiphysics analysis and optimization software	www.numeca.com
Philips	Magnotech	POC biosensor platform using magnetic nanoparticles to measure target molecules in pL samples of saliva or blood	http://www.newscenter.philips.com/main/standard/news/backgrounders/2010/20100107_magnetic_biosensor.wpd#.U3rRstKSyuI
Radiometer	AQT90 Flex	Immunoassay analyzer, determine the level of an antigen in blood. Analyze up to 30 samples/h, results are obtained after 10 min	http://www.radiometer.com/en/products/immunoassay/aqt90-flex-analyzer
Roche	CoaguChek®	Self-monitoring coagulation status; requires 1 drop of blood and provide accurate results within 1 min	http://www.coaguchek.net

(continued)

Table 1.1 (continued)

Company	Product	Description	Link
Siloam Biosciences	Trova™	Optical detection, dry reagents, results in 15–20 min for cardiac Troponin-I detection	http://siloambio.com/lifescienceproducts/point_of_care_test_poct_platform/
SIMTech Microfluidics Foundry	Microfluidic chips, pumps and interfacing kits	Polymer base microfluidic components for biomedical applications	www.simtech.a-star.edu.sg
Sophion	Qpatch	Patch clamp system, it contains microfluidic channels for optimized liquid handling; up to 30.000 compounds tested per 24 h	www.sophion.com
Symscape	OpenFlow	Computational fluid dynamics software	www.symscape.com
Phadia (Thermo Scientific)	ImmunoCAP	Point-of-care microfluidic platform for the evaluation of patients with allergy-related symptoms. Results in 20 min as colored bands	http://www.phadia.com/en-GB/5/Products/ImmunoCAP-Rapid
thinXXS	Disposable microfluidic devices	Used in applications such as diagnostics, pharmaceutical, analytical, and medical industries	www.thinxxs.com
TearLab Corp.	TearLab Osmolarity System	Quantitative tests for diagnosing and managing dry eye patients. Use only 50 nL of tear film to diagnose	http://www.tearlab.com
Tetracore	RedLine Alert™	Immunochromatographic test based on capillary action for the in vitro qualitative identification of *Bacillus anthracias*	www.tetracore.com
Trianja Technol.	Glass microfluidic components:	Glass-based solutions for use in miniaturized biological and chemical devices	www.trianja.com
Trinean	Xpose	High-speed quantification of DNA, RNA, and proteins. Includes microfluidic carrier samples (capillary intake), 1–16 samples can be loaded and stay stable up to 2 h (no evaporation)	www.trinean.com

(continued)

Table 1.1 (continued)

Company	Product	Description	Link
μFluidix	Lab-on-a-chip platforms	Plastic platforms for handling of sub-microliter amount of liquid fabricated	www.ufluidix.com
Universal Biosensors	One Touch® Verio™	Blood glucose sensor; requires 0.5 μL sample, results within seconds	http://www.universalbiosensors.com/Products/Verio-Product-Information.aspx

Table 1.2 Sources of information about microfluidics and lab-on-a-chip

Name	Type	Editor	Comment	Link
Computers and fluids	Journal	Elsevier	Publish the development of numerical methods relevant to fluid flow computations, computational analysis of flow physics and fluid interactions	http://www.sciencedirect.com.globalproxy.cvt.dk/science/journal/00457930
Lab on a chip	Journal	Royal Society of Chemistry	Publish both fabrication and applications of LOC devices	http://www.rsc.org/publishing/journals/lc
Microfluidics and nanofluidics	Journal	Springer	Publish reports or research, techniques, and applications in microfluidics	http://link.springer.com/journal/10404
Optofluidics, microfluidics and nanofluidics	Journal	De Gruyter	Publish mostly applications of microfluidics	http://www.degruyter.com/view/j/optof
Analytical chemistry	Journal	ACS	Publish mostly applications of LOC devices	http://pubs.acs.org/journal/ancham
Biosensors and bioelectronics	Journal	Elsevier	Publish mostly applications of LOC with integrated sensors, e.g., electrochem., optical sensors	http://www.sciencedirect.com/science/journal/09565663
Microelectronic eng.	Journal	Elsevier	Publish mostly on techniques for fabrication of LOC devices	http://www.sciencedirect.com/science/journal/01679317
Biomedical microdevices	Journal	Springer	Publish research in the diagnostic and therapeutic applications of LOC devices	http://link.springer.com/journal/10544
Biomicrofluidics	Journal	American Institute of Physics	Publish on theory and applications of microfluidics	http://scitation.aip.org/content/aip/journal/bmf
Journal of Micromechanics and Microengineering	Journal	Institute Of Physics (IOP)	Publish on microfabrication of LOC devices	http://iopscience.iop.org/0960-1317/
Journal of Microelectromech. Systems	Journal	IEEE	Publish on the theory, modelling, design, fabrication, assembly, and packaging of LOC and microfluidic devices	http://eds.ieee.org/journal-of-microelectromechanical-systems.html

Chips and tips	Blog	RSC	Blog to exchange ideas and solutions on common practical problems encountered in the fabrication and use of LOC devices. Include videos	http://blogs.rsc.org/chipsandtips http://www.youtube.com/user/labonachipVideos#p/u
CFD online	Forum		Computational fluid dynamics software community. Provides news, forum, blogs, online tools, books guide, and discussions forums	www.cfd-online.com
Lab on a chip and microfluidics	Scientific community		Web site to formulate questions and suggest answers related to microfluidics and LOC devices	http://www.linkedin.com/groups/Lab-on-chip-Microfluidic-Devices-713657
Lab-on-a-chip and microfluidics	Scientific community	Technology Networks	Provides information about research news, business, products, conferences, courses, and jobs related to microfluidics and LOC devices	http://www.technologynetworks.com/LOAC
FluidicMEMS	Blog		Perspectives on LOC, microfluidic and bioMEMS technology	http://fluidicmems.com

Table 1.3 Conferences on microfluidics, lab-on-a-chip, and microTAS technology

Conference	Organizer	Topics	Link
Internat. conf. on miniaturized systems for chemistry and life sciences (μTAS)	The Chemical and Biological Microsystems Society	Premier forum for reporting research results in microfluidics, microfabrication, and applications of Loc and μTAS devices	http://www.microtas2014.org/, www.cbmsociety.org
Conference on microfluidic handling systems	Freiburg University	Discuss the latest results in the field of microfluidic handling systems (pumps, mixers, valves, sensors)	http://www.mfhs2014.uni-freiburg.de
European conf. on microfluidics		Design, fabrication and theory of LOC devices	http://microfluidics2014.eu
Point-of-care diagnostics World conference	Select Biosciences	Discuss the point-of-care diagnostics field	http://selectbiosciences.com/conferences/index.aspx?conf=POCDWC2014
Lab-on-a-chip European congress	Select Biosciences	Discuss innovative developments in LOC, microfluidics, and microarrays	http://selectbiosciences.com/conferences/index.aspx?conf=LOACEC2014
Lab-on-a-chip and microarray World congress	Select Biosciences	Discuss innovative developments in LOC, microfluidics, and microarrays	http://selectbiosciences.com/conferences/index.aspx?conf=LOACWC2014

References

Araci IE, Brisk P (2014) Recent developments in microfluidic large scale integration. Curr Opin Biotechnol 25:60–68

Bassous E, Taub HH, Kuhn L (1977) Ink jet printing nozzle arrays etched in silicon. Appl Phys Lett 31:135–137

Brock DC (2011) A measure of success. Chem Herit Mag 29(1)

Chin CD, Linder V, Sia SK (2012) Commercialization of microfluidic point-of-care diagnostic devices. Lab Chip 12:2118–2134

Choi K, Ng AHC, Fobel R, Wheeler AR (2012) Digital microfluidics. Ann Rev Anal Chem 5:413–440

Duffy DC, McDonald C, Schueller OJA, Whitesides GM (1998) Rapid prototyping of microfluidic systems in poly(dimethylsiloxane). Anal Chem 70:4974–4984

Gazzaniga V (1999) Uroporphyria: some notes on its ancient historical background. Am J Nephrol 19:159–162

Gross BC, Erkal JL, Lockwood SY, Chen C, Spence DM (2014) Evaluation of 3D printing and its potential impact on biotechnology and the chemical sciences. Anal Chem 86:3240–3253

Harrison DJ, Flury K, Seiler K, Fan Z, Effenhauser CS, Manz A (1993) Micromachining a miniaturized capillary electrophoresis based chemical analysis system on a chip. Science 261:895–897

Kim SM, Lee SH, Suh KY (2008) Cell research with phisically modified microfluidic channels: a review. Lab Chip 8:1015–1023

Kitson PJ, Rosnes MH, Sans V, Dragone V, Cronin L (2012) Configurable 3D-printed millifluidic and microfluidic "lab on a chip" reactionware devices. Lab Chip 12:3267–3271

Kouba E, Wallen E, Pruthi RS (2007) Uroscopy by Hippocrates and Theophilus: prognosis versus diagnosis. J Urol 177:50–52

Lee A (2013) The third decade of microfluidics. Lab Chip 13:1660–1661

Mark D, von Stetten F, Zengerle R (2012) Microfluidic Apps for off-the-shelf instruments. Lab Chip 12:2464–2468

Martinez AW, Phillips ST, Butte MJ, Whitesides GM (2007) Patterned paper as a platform for inexpensive, low-volume, portable bioassays. Angew Chem 119:1340–1342

Moraes C, Labuz JM, Leung BM, Inoue M, Chun T, Takayama S (2013) On being the right size: scaling effects in designing a human-on-a-chip. Integr Biol 5(9):1149–1161

Pollack MG, Fair RB, Shenderov AD (2000) Electrowetting-based actuation of liquid droplets for microfluidic applications. Appl Phys Lett 77:1725–1726

Tasoglu S, Gurkan UA, Wang S, Demirci U (2013) Manipulating biological agents and cells in micro-scale volumes for applications in medicine. Chem Soc Rev 42:5788–5808

van der Meer AD, van den Berg A (2012) Organs-on-chips: breaking the in vitro impasse. Integr Biol 4:461–470

Wallis F (2000) Inventing diagnosis: Theophilus' *De urinis* in the classroom. Acta Hips Med Sci Hist Illus 20:31–73

Wittern-Sterzel R (2000) Diagnosis: the doctor and the urine glass. Lancet 354:SIV13

Xia Y, Whitesides GM (1998) Soft lithography. Annu Rev Mater Sci 28:153–184

Yetisen AK, Akram MS, Lowe CR (2013) Paper-based microfluidic point-of-care diagnostic devices. Lab Chip 13:2210–2251

Yetisen AK, Martinez-Hurtado JL, da Cruz VF, Simsekler MCE, Akram MS, Lowe CR (2014) The regulation of mobile medical applications. Lab Chip 14:833–840

Young EWK (2013) Cells, tissues, and organs on chips: challenges and opportunities for the cancer tumor microenvironment. Integr Biol 5:1096–1109

Chapter 2
Basic Microfluidics Theory

Winnie E. Svendsen

Abstract Flow in microsystems behaves very different than flow on the macroscale, i.e., the flow we are used to in our everyday life. The most obvious difference is that the chaotic turbulent flow we most often observe, e.g., rivers flowing or tap water running does not appear on the microscale. Here, the flow is more smooth and most often what we call laminar flow. Other parameters considered important on the macroscale such as inertia are insignificant on the microscale, whereas viscosity becomes extremely important. Diffusion, which on large scale is a hopeless parameter to use for transport, becomes significant on microscale. The surface of your system has to be considered more carefully as the surface to volume ratio (S/V) increases dramatically as you downscale your system. Take for example a cubic macrosystem with sides of 1 m, here $S/V = 6\ \text{m}^{-1}$, whereas for a system with sides of 1 μm the $V/S = 600{,}000\ \text{m}^{-1}$, which is a huge difference and has a large impact on flow behavior. In this chapter the basic microfluidic theory will be presented, enabling the reader to gain a comprehensive understanding of how liquids behave at the microscale, enough to be able to engage in design of micro systems and to support the theory used in other chapters in the book, but without going into the deep underlying theoretical approach.

2.1 Fluids

Unlike solids, fluids are mainly characterized by molecules spaced with a large distance apart and hence have a large degree of freedom to move, and for gasses, which also are considered fluids, the molecules are spaced even further apart. The definition of a fluid is “a material that deforms continuously under sheer stress” (Bruus 2008). In fluid mechanics the fluid is described as a continuum, where the average fluid parameters are examined, rather than the individual molecule’s parameters of in the fluid. The continuum approximation states: “The macroscopic properties of the fluid are the same as if the fluid were perfectly continuous, instead

W.E. Svendsen (✉)
DTU Nanotech, Technical University of Denmark, Kongens Lyngby 2800, Denmark
e-mail: wisv@nanotech.dtu.dk

J. Castillo-León, W.E. Svendsen (eds.), *Lab-on-a-Chip Devices and Micro-Total Analysis Systems*, DOI 10.1007/978-3-319-08687-3_2

of as in reality consisting of molecules" (Bruus 2008). In microfluidics this statement holds most of the times as properties calculated are on the length scale of 10 or more micrometers; whereas the intermolecular properties are on the nanometer scale (intermolecular distance in fluids is ~0.3 nm). So throughout this chapter the continuum approximation is used, unless otherwise stated. All through the chapter SI units will be used.

2.1.1 Important Fluidic Parameters

This section will discuss important parameters when describing fluids in general with special focus on what happens to these parameters in microsystems. The most important parameters characterizing a liquid and liquid movement are density (ρ), pressure (P), and viscosity (η). The density of a liquid is defined as mass (m) per unit volume (v) in units (kg/m^3).

$$\rho = \frac{m}{v} \tag{2.1}$$

The pressure of a liquid is only dependent on depth and is not relevant at all in closed microfluidic systems, however in open microfluidic systems, a pressure difference induced externally will have an effect on the flow of the liquid in the system. Pressure (P) is the ratio of force applied (F) to the area (A) over which that force is distributed in units kg/ms^2, often referred to as Pascal (Pa).

$$P = \frac{F}{A} \tag{2.2}$$

In microsystems filling height and reservoirs can be used to control the pressure and hence the flow rate in the system.

When you have a flow in a system, a resistance to the flow always exists. For liquids the flow resistance is mainly determined by the viscosity (η) of the fluid, which in turn is determined by the ratio of the shear stress to the shear rate, in units of kg/ms or Pa S. These parameters can be measured by a simple setup of two parallel plates with a distance l between them and filled with a liquid, as illustrated in Fig. 2.1. If we move the upper plate by a force F resulting in a velocity v of the plate, then liquid will experience a movement too. First the upper layer adjacent to the moving plate will be set into motion; this action will then be transferred by the momentum to the lower layer of liquid. In a steady state the velocities of these layers will range from v down to 0 (the layer closest to the stationary plate will have velocity 0). The force F acting on the area A of the liquid is the shear stress and the displacement of the liquid at the top plate Δx to the thickness of the film l is the shear strain ($\Delta x/l$). The shear strain per unit time t ($\Delta x/\Delta tl$) of the liquid is called the shear rate v/l

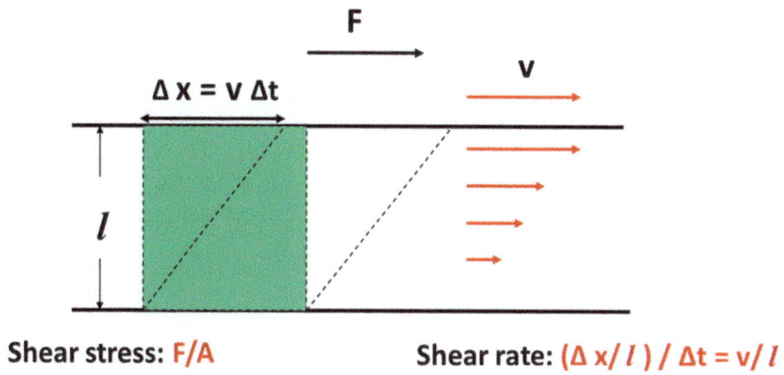

Fig. 2.1 Schematic model of viscosity

Assuming that the velocity varies linearly with position the viscosity is then given by

$$\eta = \frac{F/A}{v/l} \tag{2.3}$$

In more general terms the viscosity is given by

$$\eta = \frac{F/A}{dv/dl} \tag{2.4}$$

The above expression only holds for Newtonian liquids, i.e., where the viscosity does not depend on the fluid flow.[1] A value often used in microfluidics is the kinematic viscosity, which is the ratio of the viscosity and the density of the fluid, in unit m^2/s

$$\nu = \eta/\rho \tag{2.5}$$

With the knowledge of these simple fluid parameters we can already yield some information about the flow in microsystems. If we take the ratio of the inertial forces to the viscous forces we get a dimensionless number called the Reynolds number[2] which can be used to define the type of flow expected in the system. The Reynolds number is given by

[1] In continuum mechanics, a fluid is said to be Newtonian if the viscous stresses that arise from its flow, at every point, are proportional to the local strain rate—the rate of change of its deformation over time (e.g., water is Newtonian, while ketchup is not).

[2] Named after Osborne Reynolds (1842–1912) who popularized its use in 1833; however, the concept was introduced by George Gabriel Stokes in 1851.

$$\mathrm{Re} = \frac{\rho d v}{\eta} = \frac{d v}{\nu} \tag{2.6}$$

In this equation d is the typical length scale in the system (typically the diameter for tubes or the smallest dimension, i.e., width or height, in rectangular channels), and v is the average velocity of the moving liquid. For flow in pipes experimental studies have shown that when $\mathrm{Re} > 4{,}000$ turbulent flow occurs, and when $\mathrm{Re} < 2{,}300$ laminar flow is dominant. Between 2,300 and 4,000 the so-called transitional flow appears, where surface defects etc. can determine the flow type (Holman 1997). In a microsystem with a diameter of 100 μm, and filled with water ($\mu = 10^{-3}$ kg/ms, $\rho = 998$ kg/m^3) running through with a velocity of 1 cm/s, we get $\mathrm{Re} \approx 1$ which is far in the laminar regime. In general we can say that the flow is always laminar in microsystems. For non-circular microsystems, the hydraulic diameter D_{H} needs to be used,

$$D_{\mathrm{H}} = \frac{4A}{P} \tag{2.7}$$

Where A is the cross-sectional area and P the wetted perimeter (i.e., the total perimeter of all channel walls in contact with the liquid). With this simple discussion we can retrieve a lot of information about how fluid behaves in microsystems, however in order to analyze the dynamics of fluid flow in microsystems we need a more in-depth approach and here the Navier–Stokes formalism needs to be introduced.

2.2 Equation of Motion and the Navier–Stokes Equation

The Navier–Stokes formalism is built on the fundamental laws of conservation. We start with the equation of motion (a rewritten version of Newton's 2nd law):

$$M d_t v = \sum_j F_j \tag{2.8}$$

Where $d_t = d/d_t$ means the *total time derivative*, and we use bold symbols for vectors. We have a sum of forces $\boldsymbol{F}_j$ acting on the mass M and changing its velocity $\mathbf{v}$ in time.

When working with fluids we work with densities, i.e., divided by the volume. So expressed in *force-densities* (here we describe the motion using "*fluid-particles*") we get:

$$\rho D_t v = \rho(\partial_t v + v \cdot \nabla v) = \sum_j f_j \tag{2.9}$$

Where we now use the fluid density $\rho = \mathrm{m/V}$, and force densities $\mathbf{f}_j$, and where D_t is the *material time derivative* that can be rewritten as the operator $D_t = \partial_t + v.\ \nabla$ acting on the velocity (where $\partial_t = \partial/\partial_t$ is the *partial time derivative* and ∇ is the gradient operator). The above equation should now be imagined as the forces that are acting on a fluid-particle and thereby changing its velocity, as it flows along with the flow-field.

The two most commonly used force densities acting in microfluidic systems are the force density coming from changes in pressure p: $f_{\text{pressure}} = -\nabla p$, and the force density coming from the viscosity of the fluid: $f_{\text{viscosity}} = \eta \nabla^2 v$, where η is the viscosity of the fluid.

Putting this together we get the *Navier–Stokes equation* for *incompressible fluids*:

$$\rho(\partial_t v + v \cdot \nabla v) = -\nabla p + \eta \nabla^2 v \tag{2.10}$$

An incompressible fluid is a fluid with a density ρ that is constant in space and time, whereas a compressible fluid is a fluid with a density ρ that varies as a function of space and time.

Finally, to complete the governing equations describing the dynamics of a simple incompressible fluid, we need to include the continuity equation. The continuity equation expresses the conservation of mass: In classical fluid dynamics it states that mass can flow in and out of the volume Ω through the surface. Taking the time derivative of the mass both as a volume integral of the density as well as a surface integral of the mass current density we arrive at the continuity equation for an incompressible fluid.

$$\nabla \cdot v = 0 \tag{2.11}$$

Roughly speaking this equation says that fluid cannot arise from nowhere and cannot vanish into nowhere.

As the Navier–Stokes equation is a partial differential equation containing more unknown parameters than equations it is not possible to obtain a complete analytical solution. Due to this, it needs the conditions of all its domain boundaries to be specified in order to obtain a well-defined solution; therefore it can only be solved using specific boundary conditions and under specially defined situations. The most important and most often used boundary condition in microfluidic systems is the so-called no-slip condition, which states that the velocity at the interphases (e.g., wall/liquid) must be 0 (see section Chap. 3). An often used solution to the Navier–Stokes equation is when considering a steady, fully developed pressure-driven flow of a Newtonian fluid in a simple channel, often referred to as Poiseuille flow. The solution is called the Hagen–Poiseuille solution. In Poiseuille flow the fluid is driven through a long straight and rigid channel by imposing a pressure difference between the two ends of the channel. The velocity profile, i.e., the velocities $v(r)$ at different radial positions between the center ($r=0$) and the wall ($r=R$), in a capillary or a cylindrical channel, is given by

$$v(r) = \frac{\Delta P}{4\eta L}\left(R^2 - r^2\right) \tag{2.12}$$

Where R is the radius, L the length, η the viscosity, and ΔP the pressure drop across the channel ends, also called the hydraulic pressure. Here it is seen that the Poiseuille flow is parabolic, with the maximum velocity being in the center of the channel see figure. A parameter often used in microfluidics is the volumetric flow rate $Q = \Delta V/t = \langle v \rangle A$, where $\langle v \rangle$ is the mean velocity and A is the cross-sectional area. Given this equation this can be written as:

$$Q = \frac{\pi R^4}{8\eta L}\Delta P \tag{2.13}$$

The term $8\eta L/\pi R^4$, appearing as reciprocal in the above equation is called the hydraulic resistance (R_{Hyd}). The dependence on $1/R^4$ indicates that the fluidic resistance increases drastically as the dimensions are reduced and hence higher and higher pressure drops are needed to drive the fluid through the channels.

For a rectangular microfluidic channel, say a flat wide channel with height h and width w, where $h/w \to 0$ the flow rate is given by

$$Q = \frac{h^2 w}{12\eta L}\Delta P \tag{2.14}$$

The flow profile calculated using the no slip boundary condition for a flat wide channel with height h and width w, where $h/w \to 0$, which can be considered as an infinite long channel, is illustrated in Fig. 2.2.

Rectangular channels are the most commonly used in microfabricated microfluidic systems due to the fabrication processes used. Chapter (simulation chapter) will provide more details on the hydraulic resistance in these channels and how it can be used to obtain knowledge on the fluid transform through them.

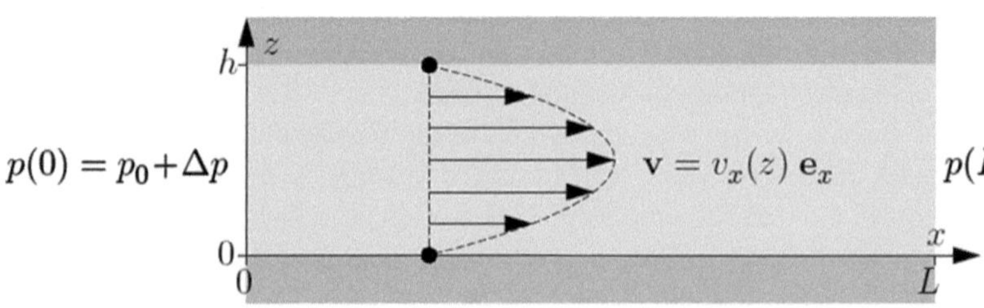

Fig. 2.2 Flow profile of fluid in a flat wide channel with height h and width w, where $h/w \to 0$, using the no slip condition

2.3 Transport Modes in Microfluidic Systems

Transport modes in microfluidic systems can be divided into direct and statistical transport mechanisms. Direct transport is where the fluid movement is controlled by applying work on the fluid inducing a volume flow with a direction and a profile. The work can be generated manually, i.e., by pressure driven flow, or electrically by electroosmotic flow. Statistical transport is entropy driven transport; this means that the fluid is more disordered after the flow than before the flow. Diffusion is a statistical transport mechanism. Generally fluid transport is not 100 % one or the other but more a combination. The more important type of transport in micro systems is advection or convection, migration, and diffusion.

2.3.1 Advection (Convection)

The terms advection and convection are often confused. Advection in chemistry and engineering is defined as: a transport mechanism of a substance, or a conserved property, by a fluid, due to the fluid's bulk motion in a particular direction, whereas convection is the combination of all flows. Both theoretical approaches can be used on liquids in this book both terms will be used depending on the field discussed, e.g., in modeling convection is always used. Free advection is when no external forces are applied. Forced advection is when fluid is moved due to an applied external force to create a directed flow. For a particle being advected (moved) by the fluid with the flow speed v, through a microfluidic system of dimension l we can introduce the advective timescale $T_C = l/v$ which is the time it takes for a particle to drift through system.

2.3.2 Migration

The directed mass transport in response to an electrical field. Electrical fields are often used to move fluid in microsystems, often the moving molecules are ionized in a polar liquid like water and hence electrical field applied to the system can direct the fluid. Although many applications use migration and the topic is important, it is not within the scope of this chapter and is not discussed further; the reader is referred to other literature like Bazant and Squires 2004, Wang et al. 2009.

2.3.3 Diffusion

Diffusion is defined as the stochastic process in which molecules drift from one region to another. Molecules move in all directions when no external forces are applied. Each molecule will move in one direction until hit by another molecule at which time it will change direction. No net flow is observable but mixing can occur. The statistical nature of movement due to diffusion can be described as random walk and the Einstein–Smoluchowski relation (here in one dimension)

$$x = \sqrt{2Dt} \tag{2.15}$$

Where x is the average distance moved by the molecule after time t in one dimension and D is a diffusion constant characteristic for the given molecule in units of (m^2/s). D expresses how fast a concentration diffuses a certain distance. Important information which can be drawn from this simple formula is that the distance a molecule travels due to diffusion is proportional to the square root of time, and hence much slower compared to directed motion which is linear with time.

To discuss the movement of fluid purely due to diffusion, Fick's law is used. Fick's law describes molecular flow or flux, where flux is defined as the rate of flow of a substance per unit area with dimensions of concentration per time per area for a chemical compound, that is, ($mol/m^2/s$). Fick's law postulates that the solute (flux) will move from a region of high concentration to a region of low concentration across a concentration gradient.

$$\partial_t c = D\nabla^2 c \tag{2.16}$$

Here c is the concentration of the solute (mol/m^3), D is the diffusion coefficient of the solute (m^2/s), and $\nabla^2 c$ is the spatial gradient of the concentration in units of (mol/m^5).

As mentioned earlier most systems are not in just the one state but in a combination. If we include advection, i.e., that the fluid velocity is nonzero, then we get the convection–diffusion equation.

$$\partial_t c + v \cdot \nabla c = D\nabla^2 c \tag{2.17}$$

From the discussion above diffusion timescale can be obtained as $T_D = l^2/D$.

2.3.4 The Péclet Number

The fact that the directed and statistical transport mechanisms are profoundly different, it is interesting to look at the relative rate of these in the microfluidic system. This can be done by writing the convection–diffusion equation in a dimensionless form. When you do this a constant called the Péclet number is obtained.[3] The Péclet number is defined as the ratio between the advective and diffusive transport and is given by:

$$\mathrm{Pe} = \frac{\text{Advectivetransport}}{\text{Diffusivetransport}} = \frac{lv}{D} \tag{2.18}$$

Where v and l are the characteristic velocity and characteristic length of the system (in this case the length over which the advection and diffusion are going to take place) and D is the diffusion coefficient of the species that is being transported.

As the Péclet number for a given system becomes larger, diffusion becomes less and less important. It is general to say that if the Péclet number is over 1,000 then the diffusion of the solute can be neglected. If the Péclet number is less than 10 then the diffusion process dominates and the advection can be neglected.

To visualize the transport processes in microfluidic system let's look at some examples where we assume that the distance to be travelled is 100 μm and the velocity of the fluid is 1 cm/s giving an advection time of 10 ms.

Species (in water)	Diffusion constant D (m^2/s)	T_D (s)	Pe
Small ions	2×10^{-9}	5	500
30 base pair DNA molecules	4×10^{-11}	250	25,000
5,000 base pair DNA molecules	4×10^{-12}	10,000	250,000
A cell or larger molecules (20 μm in diameter)	2×10^{-14}	500,000	50,000,000

So for small ions diffusion is relatively fast and may interfere with the advective transport of these but if you want to look at interactions of larger species you can end up waiting a long time if you only rely on diffusive processes. From the above table we can also see where the limit of a Péclet number of 1,000 to neglect diffusion comes from. The 10 ms of advection time are relatively close to the 5 s of diffusion time in the case where the Péclet number is under 1,000. In all other cases diffusion can be neglected as no diffusive transport will be observed by the time the particle has left the channel.

2.4 Additional Requirements to Fully Model a Flow System

In the next chapter examples of modeling microfluidic systems are given; some notes are added here which you have to take into account when using the above theory to do modeling. In order to reduce computational time a model region and boundary conditions have to be defined.

[3] Named after the French physicist Jean Claude Eugène Péclet (1793–1857).

2.4.1 Model Region Ω

Now with the Navier–Stokes equation, we can describe how fluid-particles interact mutually due to viscosity and how that can be driven by a pressure difference.

Still we need to define at what region the fluid flows, and that is called the model region, denoted Ω.

In reality the region is three-dimensional (3D), but often, the modeled flow does not vary along a given direction, and this makes it possible to simplify the model (reduce computational time) to 2D or sometimes 1D (flow in tube when expressed in polar coordinates).

2.4.2 Boundary Conditions

It is not enough to describe the region—also what happens on the boundary of the region is very important! The boundary of the region Ω is denoted ∂Ω.

Simple flow systems only need one type of boundary condition, the so-called no-slip boundary condition, which states that for viscous fluids at a solid boundary, the fluid will have zero velocity relative to the boundary.

$$v(r) = v_{\text{wall}}, \quad \text{for}\, \boldsymbol{r} \in \partial\Omega \tag{2.19}$$

You also need to specify where your fluid comes in and out of the system as well as with what force it moves. This can be specified as a pressure or a velocity.

References

Bazant MZ, Squires T (2004) Induced-charge electrokinetic phenomena: theory and microfluidic applications. Phys Rev Lett 92(6):661011–661014

Bruus H (2008) Theoretical microfluidics. Oxford University Press, Oxford, UK

Holman JP (1997) Principles of convection. In: Heat transfer. McGraw-Hill, Boston, USA

Wang X, Cheng C, Wang S, Liu S (2009) Electroosmotic pumps and their applications in microfluidic systems. Microfluid Nanofluid 6:145–162

Chapter 3
Design and Simulation of Lab-on-a-Chip Devices

Maria Dimaki and Fridolin Okkels

Abstract Microfluidic channels are an essential part of any lab-on-a-chip system. They usually perform various functions, such as transporting liquids from A to B or mixing or separating liquids. As production costs for such systems are not insignificant, it is essential that the systems are designed properly before the fabrication, in order to avoid unnecessary fabrication repetitions. The use of simulations can give a good idea of how microfluidic systems work, to the point where a significant part of the design optimisation can be done theoretically. This chapter will provide some basic information on how to embark on these types of simulations, explaining the basics of microfluidic modelling and providing examples.

The development of lab-on-a-chip systems relies heavily on the presence and correct function of microfluidic channels and other liquid handling components that can play various roles: They can be there merely to transport liquids from A to B, or they can have more active roles, i.e., be structured in a way that can help mix two fluids, have structures that apply forces on particles, e.g., electrodes for electrical forces in order to sort particles etc. Especially in the case where microfluidic channels are needed for something more than transport, it is an advantage to have an idea how they are going to work before embarking on expensive and time-consuming production of the structures.

A very inexpensive, but not necessarily fast way of ensuring that your envisioned structures are indeed going to do what you want them to do is to simulate their function using analytical or—most often—numerical calculations. Optimization of the design can be obtained in such a way before the fabrication process and its function can be tested under several working conditions.

Although simulations provide valuable input, one should keep in mind that a simulation is only as accurate as you set it up to be. A lot of assumptions are being made when a real-life problem is put on paper, and these can in principle be wrong or at least not very good. Also, even though we can try and simulate all the effects that can take place when we do an experiment, this is in practice quite difficult.

M. Dimaki (✉) • F. Okkels
Department of Micro- and Nanotechnology, Technical University of Denmark, Bldg. 345E, 2800, Kgs. Lyngby, Denmark
e-mail: maria.dimaki@nanotech.dtu.dk

J. Castillo-León, W.E. Svendsen (eds.), *Lab-on-a-Chip Devices and Micro-Total Analysis Systems*, DOI 10.1007/978-3-319-08687-3_3

Take for example a situation where a voltage is applied in electrodes within a microfluidic channel in order to affect charged particles that flow within. You can easily calculate the force applied to the particles, as well as the particles' trajectories. You can even simulate how the voltage will influence the liquid due to electroosmotic phenomena in some but not all cases. But there is a number of other things that could take place and which are difficult to simulate, mainly because there is no accurate model or no knowledge of the values of the parameters involved, for example electrothermal forces. For that reason you should not use all your time doing simulations. It is difficult to provide a number, as it depends on the application, but generally not more than 30 % of the total development time should be used on simulations.

This chapter will provide information about how to set up a microfluidic simulation, concentrating mainly on how to simulate flow through a channel, as well as species transport within a channel. The chapter will touch subjects such as dimensionality of the problem, i.e., simulation in 1D, 2D, etc, concepts such as the Reynolds and Péclet numbers, lumped elements for faster calculations, mesh setup and importance of a good mesh, and simulations of species diffusion and mixing. Microfluidic theory is provided in Chap. 2; therefore only little theory is going to be presented here.

A number of finite element modeling software is available, all having their advantages and disadvantages. It is not the aim of this chapter to provide recipes for one or more of these; therefore the concepts that are being presented are generalized and can be applied on any of these simulation tools. The examples presented in this chapter have all been done using COMSOL multiphysics, which is the software used by the authors currently. At the end of the chapter you will find a list with software that can be used along with a brief discussion. You can also find some interesting papers dealing with simulations of microfluidic systems.

3.1 Setting Up a Simulation: A Quick Checklist

In recent years, modern simulation software has become very user friendly and provides more or less ready solutions to almost every conceivable physics problem. A lot of the available software has preset equations to calculate not only flow, but also dielectrophoretic forces, gravitational forces, turbulent flow, etc. Quite a few can now also per default provide a decent mesh so that all you need to do is design your geometry, choose the materials and press start. Although this approach works for a lot of simple geometries and problems, it is not advisable, as one often ends up with a large geometry with a large number of elements that takes hours to solve.

Instead, one should always examine the problem at hand. Is a simulation necessary? Why? What information will come out of the simulation, which was not easily achievable by simple analytical calculations? If we establish that a numerical simulation is necessary, then the next step is to figure out what kind of problem we want to solve. Is it a pure flow problem, e.g., how the flow will look like in a complicated (perhaps 3D) geometry? Or do we also need to introduce species that can diffuse, e.g., when we want to see how a flow from an inlet can be focused by being pressed by flows from other inlets, or how diffusion works in bends? Are there other physical phenomena affecting the flow, e.g., a heat source, a moving structure or similar? Answers to these questions help us establish what kind of physics we need to include in our simulation and how the different physics are connected with each other.

This latter question, how physics are connected with each other, can save us a lot of time and computational power. Take for example the problem where two inlets merge into one channel. One of the inlets contains some species of concentration c, while the other does not. We are interested to see how the species will mix with the non-species carrying fluid in the common channel. We need to solve two problems: the actual flow of fluid in the channel and the convection and diffusion of species. Although the two problems can be solved simultaneously, there is no reason for it. The flow profile will look the same, regardless of the concentration c. However, the diffusion of the species will depend on the flow profile. This type of problems are called one-way coupled and the solution time can be greatly reduced by first solving the flow problem and then using the calculated velocity profile as input to solving the convection/diffusion problem. Should we need to change the concentration of the species, then we do not need to re-solve for the velocity, but only for the convection/diffusion problem. We use this example more in the coming sections to illustrate the various points.

When the above issues are resolved then it is time to look on the specifics of the geometry. Clearly, in real life everything is in 3D. However, a few times it is possible to take advantage of symmetries or invariability in one axis to simplify the geometry to 2D or even 1D. Doing this greatly reduces the computations and the times needed for the simulation, however, it should be applied with care. Sect. 3.4 presents details of this dimension analysis, and gives some tips and tricks on what to look for and how to apply these shortcuts.

Another geometry issue is how much of your structure you need to simulate. Let us return to the example of the two inlets that merge into a channel. Undoubtedly in practice our two inlet channels will be a few mm long before they merge and the outlet will also most likely be a few mm away from the merging region. However, modeling this many mm long system is not necessary at all. The interesting part for modeling is where the two inlet channels meet and merge and then some length inside the common channel where mixing takes place. Simple analytical calculations taking velocity and diffusion into account will provide us with an order of magnitude estimate on the length of the common channel that we need to model in order to see what is happening. The rest is just "dead space," a part of the channel where the fluid just flows with the velocity specified. These parts can easily be replaced by the so-called lumped elements, which are described further in Sect. 3.3.

Finally, after you have set up your geometry, your physics, boundary conditions, you are ready to solve your problem. To do so you need a mesh. The quality of your mesh is directly proportional to the accuracy of your solution. A bad mesh will give you an answer, too, but whether or not that answer is accurate is another story. Mesh generation is greatly dependent on the software you use and there are many ways different software improves the mesh. But some general concepts and practices will be presented in Sect. 3.7.

3.2 How to Approach Microfluidics Modeling

Even for very simple microfluidic systems that are very easy to model in commercial simulation software packages such as COMSOL, a too hasty numerical modeling approach can lead to false or even completely misleading results. Therefore we are encouraging all students, teachers, and researchers that are planning to set up microfluidic systems, and who want to base their design on more than "Trial and error," to consider how to proceed.

Should you go straight to the computer and start the simulation software or should you rather fetch a pen and an empty sheet of paper?

We will argue about this choice in the following.

3.2.1 The Wonders of Present Simulation Software

Within the last 10 years, simulation software packages have improved greatly! Earlier on, each element used in the modeling process, such as geometrical design, mesh, numerical solvers, and visualization should be created or performed in separate programs, and much effort laid in managing and transferring the different model elements between the required programs. The resulting simulation procedure was therefore very complex and required training and experience.

Now all the elements of a complete numerical modeling process are covered and combined in a single software, seamlessly combining all the steps from building the geometry, adding the involved physical properties and dynamics, solving the model numerically to analyzing, visualizing and exporting the final results. This greatly reduces the complexity of working with numerical modeling, and therefore their use has also become "Mainstream," reaching out to a large group of users. This evolution is a great success for the software developers, and has boosted the use of numerical modeling within all branches of science and industry, but there is also the following catch: One might say that these programs have become too easy to use, such that you replace the actual modeling process of your system with the fixed building process supplied by the numerical modeling program. You are thereby limiting the whole process to that of the capabilities of the program.

Many essential properties of the planned system can be estimated theoretically within 10 % accuracy, which is well enough for judging if the chosen approach will fulfill the requirements originally set for the system. Moreover, the time used on achieving the theoretical estimate is often an order of magnitude smaller than the

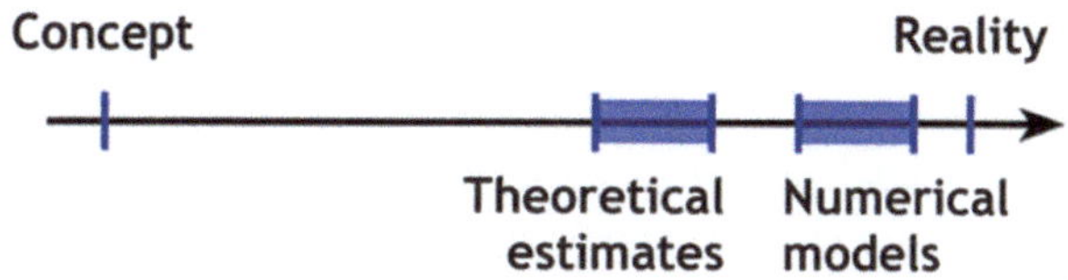

Fig. 3.1 "Distance to reality" of different modeling approaches, with the initial concept without any insight to the *left*, and full insight, i.e., coming from the experimental results to the *right*

time needed to set up the numerical model. If your primary goal, during the early stage of system design, is to investigate the durability of the proposed design, then fast initial theoretical estimates will enable you to consider many alternative candidate solutions; at the same time it would require to set up one single numerical model of your first candidate solution, and make one realization for some given system-parameters.

Below in Fig. 3.1 we illustrate our view on how far both the theoretical and numerical approach can bring us towards getting insight into a given proposed system. We imagine an "insight scale-bar" or "distance to realism," where to the far left, we have the pure concept of the system. This could be obtained by answering questions like: "Which physical and/or chemical effect do we utilize?", "What responses are we looking for?", and "How should a rough candidate design with the involved elements look like?" In the other end of the scale the ultimate insight into the system would be the actual experimental results, which define the far right of the scale, and while this often is the final goal, the idea of modeling systems is indeed to save time and resources in getting valuable insight into the different candidate systems before actually conducting the often laborious experimental work.

There is no doubt that good numerical models are superior in predicting the system responses, and thereby can reach very close to the insight gained from performing the real experiment. Therefore the range of numerical models spans an interval from close to experimental verification and a bit downward, since some systems involve very complex properties that are intrinsically hard to model numerically such that they will limit your insight obtained even from state-of-the-art numerical models. We state that well-chosen theoretical estimates can reach quite close to the level of insight given by numerical models, where again the applicability may vary depending on the type of system, as seen in Fig. 3.1. Since such theoretical estimates can be quite easy to perform, we therefore encourage all to use the essential insight that can be gained from simple theoretical models as a first approach to system design. In the following we argue for pros and cons of applying theoretical models early in the system design phase.

3.2.2 Theoretical Versus Numerical Modeling

We give in the following table our version of the pros and cons of theoretical versus numerical modeling in the early system design phase. For each kind of activity related to modeling, we compare it between "Theoretical estimates" and "Numerical simulations" by first giving a small headline, followed by a rating within the following interval of choices [(−),(~),(+),(++)], where (~) represents neutral/"either-or". Then we give a small argument for each choice.

3.2.3 The Strength of Dimensionless Numbers

In Table 3.1, we had to reserve a double-plus (++) in rating the use of dimensionless numbers as a theoretical tool—due to its easy generation of valuable information. This information comes by comparing the mutual influence of different properties in the system.

The simple reason that these numbers are dimensionless is that they evaluate the ratio between similar quantities, such that the most dominating forces acting in the system, or two important time-scales.

We will focus on the Reynolds number and the Péclet number, which are the most obvious choices for Lab-on-a-Chip systems, taken out of a broad palette of dimensionless numbers that are valuable in other types of systems.[1]

3.2.4 The Reynolds Number

In the simplest flow systems, described only by the flow velocity and the pressure, only one dimensionless quantity is needed, namely the Reynolds number (Re) that evaluates the ratio between the characteristic inertial force and the characteristic viscous force in the system, Eq. (2.6).

The Reynolds number can be quantized by the expressions above by rewriting the governing equation for fluid flow (the Navier–Stokes equation, see Sect. 2.2) using a characteristic length-scale L and a characteristic flow speed U of the system, with the fluid density ρ, the dynamic viscosity η, or using the kinematic viscosity ν. Normally L is chosen as a small typical length-scale in the flow system, and U as the mean flow speed.

For small values of the Reynolds number, $\mathrm{Re} < 1$, the viscous forces completely dominate the flow, and the flow is called laminar. Laminar flows are the steady smooth flows that are well-known from the handling of everyday thick, viscous

[1] See for example Appendix B in "Theoretical Microfluidics" by H. Bruus, Oxford University Press, 2008.

Table 3.1 Pros and cons of theoretical estimates and numerical modeling

Activity	Theoretical estimates	Numerical simulation
Preparation	Learn the theory: (+) Admitted, it takes a considerable effort to learn to work with theoretical modeling, but the time is returned in multitude, as it not only gives fast theoretical estimates, but also gives a better insight into the whole system	The Software modeling guide: (~) Modern simulation software guide the user through the modeling process, and this has drastically reduced the required preceding knowledge on the modeling process. This both broadens the use of such software, but also gives the user a false security in relying on the produced results. (Many believe that the program cannot make wrong results)
Output (mostly visual)	Formulas look dry: (−) No matter how useful the final algebraic results are for gaining insight into the system, the formulas need to be interpreted, and this discourages many from obtaining the valuable information they contain	Nice colorful visualizations: (+) The trademark of numerical modeling has always been impressive visualizations of the results, and the colorful surfaces or breathtaking 3D illustrations have through years supported the general confidence in numerical models as an indisputable truth—but the quality of the output is only as good as the quality of the input!
Applicability range	Approximate, but broad ranging: (+) In order for the analytical model to be manageable, the involved system often has to be hugely simplified. Still analytical results have the benefit that they directly relate the different parameters in the model, such that the impact of varying a given parameter easily can be examined	Only one specific solution at the time: (−) Even with today's software, only a single parameter choice can be evaluated at the time. You can then scan through some of the parameter space, or use parallel computing, but the overall dependence on the parameters have to be assembled from a limited set of single realizations
Insight in the physics	Dimensionless numbers: (++) The insight from quickly calculating the relevant dimensionless number (s) cannot be overestimated! Knowing such numbers can easily determine the importance of different physical properties such as if different substances will mix spontaneously by diffusion (Péclet number)	Better details: (+) When the complexity of the model geometry increases, there is no way around numerical models, and if they are built properly, the results gained on for example the spatial distribution and/or temporal dynamics of a given field variable are invaluable
Implementation time	Fast: (+) Knowing the limitations laying in the theoretical modeling approach, and therefore what kind answers are obtainable, the actual calculations are durable and quickly obtained. The time you use is well spent compared to the insight you gain	Still takes time: (−) The modeling process implemented in the different simulation software has improved, and this is partly because the developers have made a lot of predefined choices for the user. Still the amounts of time you spend on building the model often exceeds the insight gained from the final results

liquids such as mayonnaise or liquid honey. This is in complete contrast to high Reynolds number fluids, where the inertial force, i.e., the weight of the fluid, dominates the motion, and such flows are called turbulent. Strong turbulent flows are seen in for example the atmosphere or violent flowing rivers, and are dominated by a complex, irregular motion of eddies on all length-scales.

Because of the sub-millimeter length-scale of channels in Lab-on-a-Chip systems, the corresponding Reynolds numbers are always low, and we can assume that the microfluidic flows are laminar. This is very handy when modeling such flows, as no spontaneous irregularities or eddies thereby occur in normal micro-channels. This makes it possible to use simple linear flow-relations as the Lumped Element modeling, presented in one of the following sections. Still you should always check if the Reynolds number is indeed below unity, before you apply such linear models.

3.2.5 Péclet Number

Once we introduce different molecular species, and thereby varying concentration fields, there are mainly two kind of transport inside Lab-on-a-Chip systems:

- *Advective transport*, where the molecules follow the flow of the surrounding fluid.
- *Diffusive transport*, that can either consist of Brownian motion for medium sized particles, or a diffusive flux, proportional to the concentration gradient (Fick's law) for substances described by a related concentration field.

Whereas advective transport by laminar flows tends to create strong concentration gradients, diffusive transport smears out concentration variation. Therefore their mutual ratio gives valuable information about the general transport properties of the species, and this leads directly to the definition of the Péclet number (Pé), Eq. (2.18).

Generally, the species transport in systems characterized by small Péclet numbers (Pé $<$ 10) will be governed by diffusion, and concentration gradients will spontaneously spread out, independently of the underlying flow pattern. On the contrary, species transport in systems characterized by large Péclet numbers (Pé $>$ 1,000) will be strongly governed by the flow pattern, where strong concentration gradients will persist through the system. The effect of varying the Péclet number on a numerical model of how two miscible fluids merge downstream are shown below in Fig. 3.2. A side fluid inlet containing a diffusive substance is added to a main fluid flow, and the transition from low Pe diffusive transport to high Pe advective transport is clearly illustrated.

Knowing the three characteristic quantities (L, U, and D) of a fluidic system, the Péclet number can be computed with minimal effort, and knowing its value gives direct information about the mixing properties of the system. This will then help in deciding if diffusion does the job, or if additional means have to be taken to ensure good mixing.

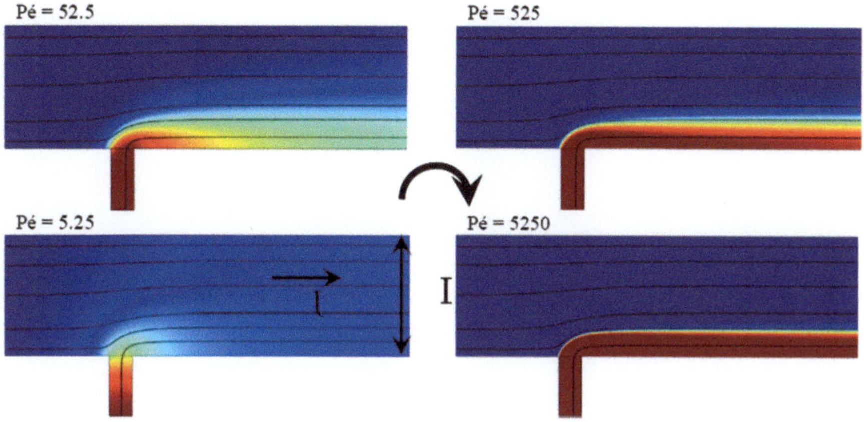

Fig. 3.2 Numerical example, showing the influence of varying Péclet numbers. A side fluid inlet containing a diffusive substance is added to a main fluid flow, and the transition from diffusive transport to advective transport is clearly illustrated, as the Péclet number for the different examples increases in steps of one order of magnitudes. (Simulations done in COMSOL by the authors)

In that way, dimensionless quantities give quick and important estimates, valuable for all researchers within a large group of natural and technical sciences.

3.3 Lumped Element Modeling

One of the biggest issues when modeling real systems is that the size of these is usually forbidding for finite elements simulations. Microfluidic channels are often long from inlet to outlet, however, usually only a smaller part is the "active" part of the system, which needs to be simulated. For the remaining parts basic microfluidic calculations can be applied, e.g., in order to replace the long inlet and outlet channels with equivalent hydraulic resistances. The following section will present when such a replacement is possible and how it can be calculated. An example of how a simulation result changes when this is used will be given.

3.3.1 Hydraulic Resistance

For a pressure-driven, steady-state flow of an incompressible Newtonian fluid through a straight channel (the so-called Poiseuille flow), and a constant pressure drop Δp will result in a constant flow rate Q. The proportionality factor of this relationship is called the hydraulic resistance of the channel, so that we can write the Hagen–Poiseuille law as:

$$\Delta p = R_{\text{hyd}} Q \tag{3.1}$$

The hydraulic resistance has thus units of kg/m^{4}s.

In the majority of the cases the fabricated microfluidic channels have a rectangular cross-section. The hydraulic resistance in this case is given by

$$R_{\text{hyd}} = \frac{12\eta L}{1 - 0.63(h/w)} \frac{1}{h^3 w} \tag{3.2}$$

Where η is the viscosity of the fluid, L the length, h the height and w the width of the microfluidic channel, with $h < w$. Other solutions to the Navier–Stokes equation can be obtained for other channel cross-sections.[2]

An important note is that Eq. (3.2) can only be derived if the nonlinear term in the Navier–Stokes equation can be neglected. This is only the case when the Reynolds number of the channel is much smaller than 1.

The hydraulic resistance is completely analogous to the electrical resistance. Therefore it can also be used in the same way, when we have straight channels connected in series or in parallel. However, one should always make sure that the Reynolds number is smaller than 1 before using these equations.

Serial and Parallel Coupling of Straight Channels

Consider two straight channels as shown in Fig. 3.3 that are connected in series or in parallel. In the case of the serial coupling the pressure drops across the two channels can be added to each other (as in the case of adding voltages in the case of two electrical resistors in series), so that the total pressure drop of the coupled channels is given by:

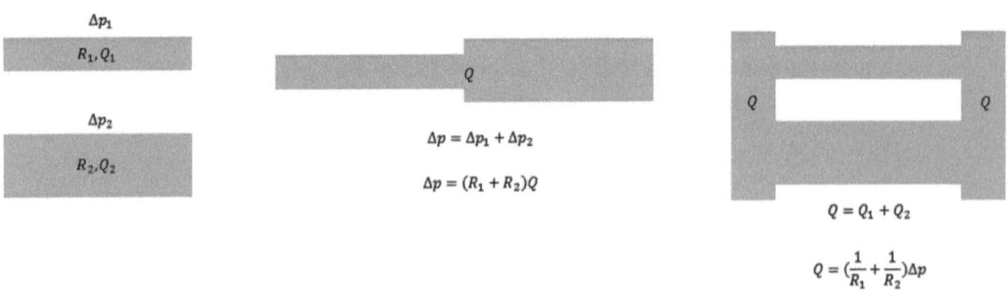

Fig. 3.3 Channels connected in series and in parallel

[2] See Chap. 4 in "Theoretical Microfluidics" by H. Bruus, Oxford University Press, 2008.

$$\begin{matrix} \Delta p_1 = R_1 Q_1 \\ \Delta p_2 = R_2 Q_2 \end{matrix} \rightarrow \Delta \mathrm{p} = (R_1 + R_2)Q \rightarrow R_{\mathrm{tot}} = R_1 + R_2 \tag{3.3}$$

Therefore the total hydraulic resistance of the system is simply the sum of the hydraulic resistances of the two connected channels.

In the case of the parallel coupling of the two channels it is the conservation of the flow rate that applies, so that

$$Q = Q_1 + Q_2 \Rightarrow Q = \left(\frac{1}{R_1} + \frac{1}{R_2}\right)\Delta \mathrm{p} \rightarrow R_{\mathrm{tot}} = \frac{R_1 R_2}{R_1 + R_2} \tag{3.4}$$

Example

To demonstrate the importance of the hydraulic resistance, let us take an example of a channel that is splitting up in two outlet channels (called channel 1 and channel 2) after a certain length L. It is our intention to split the flow into two equal parts in the two outlet channels. Assuming that the lengths of these two channels are the same, this will also be the case. However, in practice, it is relatively difficult to get the lengths to be exactly the same, mainly due to fabrication uncertainties but also due to the various tubing used to connect the microfluidic channels to the outside equipment. It is important to know how a small variation in the lengths (or widths or heights) of the outlet channels can influence the flow separation. We assume that the pressure at both outlets is the same (atmospheric pressure). From Eq. (3.2) we can see that the hydraulic resistance is proportional to the length. Assume that channel 2 has a 1 % longer length than channel 1, i.e., $L_2 = 1.01L_1$. Then $R_2 = 1.01R_1$ and consequently $Q_2 \approx 0.99Q_1$, which means that the flow will not be divided equally between the two outlets.

If we now decide that we would like to obtain a particular flow split ratio, other than 1, and then we can back-calculate the required ratio of the channel hydraulic resistances and use that in our simulations without having to model our entire system. Different simulation programs achieve this in different ways, but for example in COMSOL Multiphysics one can use the concept of "laminar inflow" and "laminar outflow" in order to provide the length of the channel that is after the boundary defined as outlet. By simply changing this length the flow division in the two outlet channels will change without us having to use a different geometry and a new mesh, as shown in Fig. 3.4.

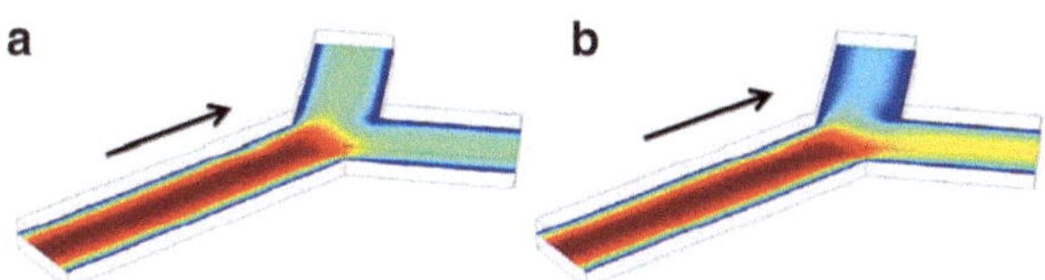

Fig. 3.4 (**a**) Velocity profile inside the channel at the splitting region when the hydraulic resistance ratio in the two outlets is 1:1. (**b**) Velocity profile when the hydraulic resistance ratio is 2:1

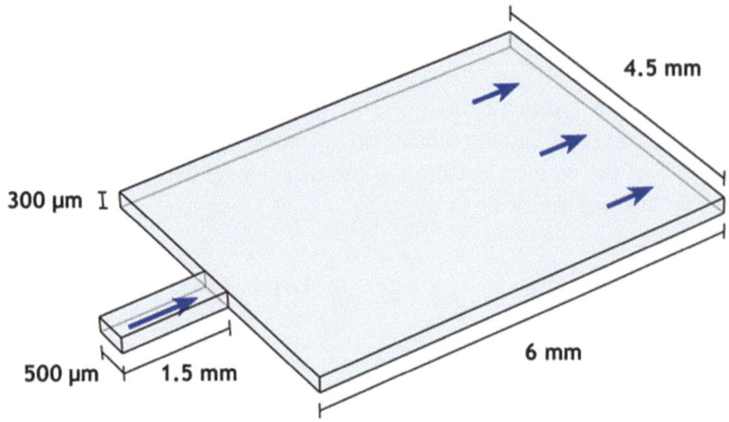

Fig. 3.5 Illustration of the simple channel setup with specification of dimensions and flow *arrows*, used in the following to exemplify the different modeling approaches

3.4 Illustration of Dimensional Analysis by Simple Microfluidics Example

The following example is nearly as simple as it can get, but has nevertheless caused troubles for some of our former students in a microfluidics modeling course: They had to model the flow of water from a short narrow microchannel to a wider longer microchannel. The full layout is shown below in Fig. 3.5 with specification of dimensions, and the height is the same for the two channels.

Pure water is flowing through the system, so we only need to gain insight in the flow-field through the channels. The flow is driven by a pressure drop of 1 Pa (Pascal) between inlet and outlet.

3.4.1 Direct Three-Dimensional Numerical Modeling

Since the modeling of both the channel geometry and the fluid dynamics is a simple task, a direct numerical simulation of the full 3D flow-field is quickly made (with the aid of efficient simulation software) and 3D streamlines of the flow are shown in Fig. 3.6. This once and for all gives all the information needed about this simple system. Still we can utilize some properties of the system to simplify the needed model, as shown in the following.

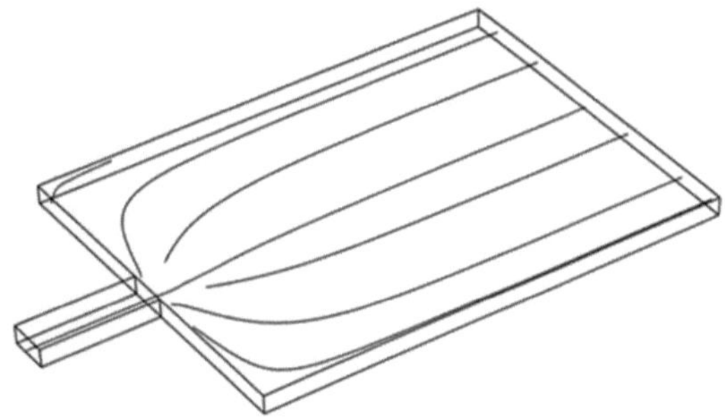

Fig. 3.6 Direct numerical simulation of the full 3D flow field of the given simple example. (Simulations done in COMSOL by the authors)

3.4.2 Simplification by Using a Quasi-3D Flow-Model

Since in this example, the channel height is constant, it is advantageous to model the system with a quasi-3D flow-model. As the name describes, it models the 3D flow by solving a modified flow problem in the 2D plane by averaging the velocity variations in the height dimension. To be more specific, the viscous damping of the fluid from the top and button of the channels is implemented in the quasi-3D model by an additional damping force term, denoted the Darcy damping term:

$$\vec{F}_{\mathrm{Da}} = -\alpha \vec{u}, \quad \text{with the damping coefficient}\, \alpha = \frac{12\,\eta}{h^2} \tag{3.5}$$

As seen above, the damping coefficient α depends only on the viscosity η and the channel height h (the direction over which the velocity is averaged), and for this quasi-3D flow-model to be accurate in the given implementation, the height should be constant, as in the case in the example. The resulting 2D flow-field is shown in Fig. 3.7 visualized by streamlines and the flow speed in color-coding.

Comparing the streamline pattern of the full 3D simulation in Fig. 3.6 and the quasi-3D flow-model in Fig. 3.7 shows that indeed the quasi-3D is capable of capturing the overall flow-pattern of the system.

3.4.3 Theoretical Estimations

Our last modeling approach is theoretical, and in this case, the applied equivalent circuit model is particularly simple. This is because the hydraulic resistance only has to be calculated for two channels, which are placed in series connection, such that the resistances just have to be added. Knowing the pressure drop, the flow-rate can be calculated, and from there the mean velocity can be estimated in the broad channel.

For the first narrow channel, the width is comparable to the height, and therefore the extended relation has to be used:

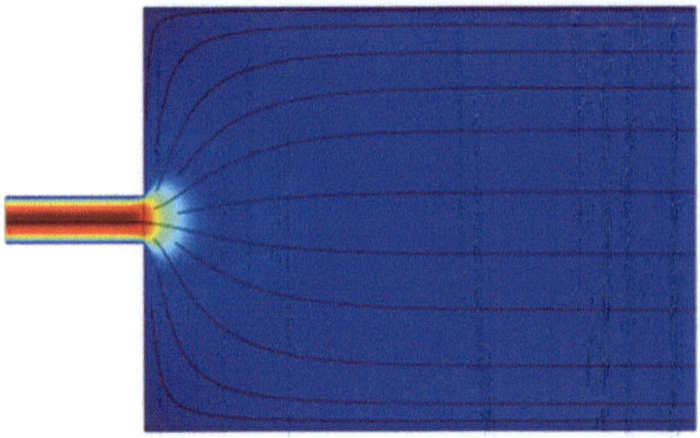

Fig. 3.7 Flow field calculated using quasi-3D flow model and visualized by streamlines and the flow speed in color-coding

$$R_{\mathrm{Hyd},1} = \frac{12\eta L}{[1-0.63(h/w)]h^3 w} = \frac{12\times10^{-3}\mathrm{Pa\,s}\;1.5\times10^{-3}\,\mathrm{m}}{\left[1-0.63\left(\frac{3}{5}\right)\right]\left(3\times10^{-4}\,\mathrm{m}\right)^3\;5\times10^{-4}\,\mathrm{m}}$$
$$= 2.14\times10^{9}\ \mathrm{Pa\,s/m^3} \tag{3.6}$$

where we have omitted the units when entering the values for the *h*/*w* ratio.

For the second channel, the correction-term in the extended relation can be omitted as the width is much larger than the height, and we get:

$$R_{\mathrm{Hyd},2} = \frac{12\eta L}{h^3 w} = \frac{12\ \times10^{-3}\,\mathrm{Pa\,s}\;6\times10^{-3}\,\mathrm{m}}{\left(3\times10^{-4}\,\mathrm{m}\right)^3\;4.5\times10^{-3}\,\mathrm{m}} = 0.593\times10^{9}\ \mathrm{Pa\ s/m^3} \tag{3.7}$$

We have kept the result in units of $10^9\,\mathrm{Pa\ s/m^3}$ to emphasize the much larger contribution from the smaller channel than from the larger, wider channel.

Driving the fluid with a pressure difference of $\Delta p = 1\,\mathrm{Pa}$ we apply the Hagen–Poiseuille law to calculate the total flow rate, where the total hydraulic resistance is just the sum of the contribution from each channel:

$$\Delta p = \left(R_{\mathrm{hyd},1} + R_{\mathrm{hyd},2}\right) Q \tag{3.8}$$

giving $Q = 3.65\times10^{-8}\,\mathrm{m^3/s} = 11.0\,\mu\mathrm{L/min}$ that relates to an approximate outlet velocity of

$$u = \frac{Q}{w\,h} = \frac{3.65\times10^{-8}\,\mathrm{m^3/s}}{5\times10^{-4}\,\mathrm{m}\;3\times10^{-4}\,\mathrm{m}} = 2.71\times10^{-4}\,\mathrm{m/s} = 0.271\,\mathrm{mm/s} \tag{3.9}$$

As a last added bonus, we can calculate the Reynolds number:

$$\mathrm{Re} = \frac{\rho\, U\, L}{\eta} = \frac{10^3\,\mathrm{kg/m^3}\;\, 2.71 \times 10^{-4}\,\mathrm{m/s}\;\, 3 \times 10^{-4}\,\mathrm{m}}{10^{-3}\,\mathrm{Pa\,s}} = 0.081 \tag{3.10}$$

which is much smaller than unity, meaning that it is a creeping flow and it is valid to apply the equivalent circuit model.

3.4.4 Comparing the Different Approaches

The best way to evaluate the different modeling approaches of the example is to plot the flow profile close to the broad outlet, as shown in Fig. 3.8. In this case we have averaged over the height direction for the full 3D model (solid blue curve), and show the flow profile as a function of the distance away from the side wall. This is exactly the field calculated in the quasi-3D flow-model (solid red curve), and this enables a direct comparison, where both curves go to zero at the wall. To also compare the theoretical estimation, we have added the estimated mean flow speed (dashed blue line) which shows that a simple linear theoretical model can predict the result within 10 %. The black vertical dashed line illustrates the good "rule of thumb" that the impact from the sidewalls on the averaged velocity extends only a length-scale into the channel, approximately equal to the height of the channel.

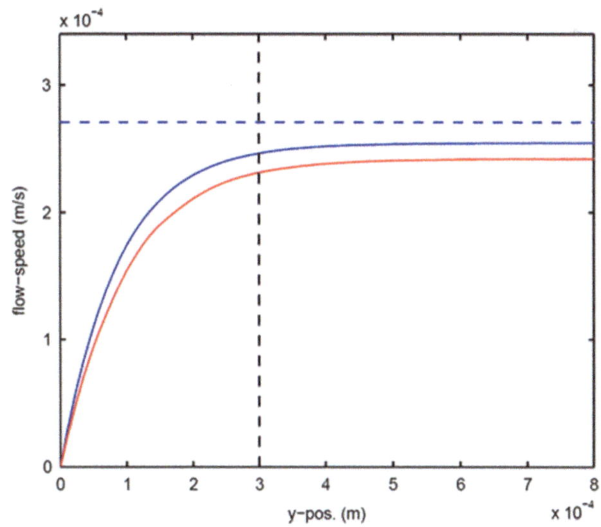

Fig. 3.8 Comparison of flow profiles averaged over the height direction. The full 3D model (*solid blue*), the quasi-3D model (*solid red*), and the theoretical model (*dashed blue*). The *vertical dashed line* illustrates the "rule of thumb" that the impact from the sidewalls on the averaged velocity extends only a length approximately equal to the height of the channel

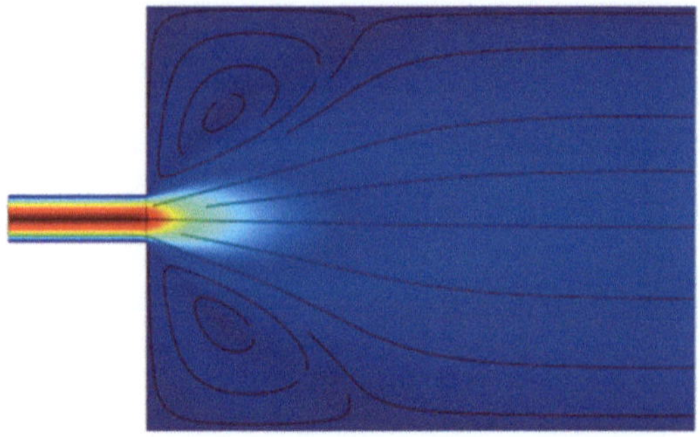

Fig. 3.9 The flow-field of the fatal choice of working with a pure 2D model, ignoring the contribution from the channel height. Visualized in the same way as Fig. 3.7

3.5 When Doing Numerical Modeling: Do It Right!

We introduced this simple example because it played a trick on some students, and here is the reason why: Observing that the simple channel geometry only changes its width and length, one might be tempted to focus on these variations only, and ignoring the height of the channels. Thereby one could choose to make a pure 2D model of the given example, and that is indeed a total failure!—As will be illustrated in the following.

The resulting velocity-field of the pure 2D flow-model is shown in Fig. 3.9, and visualizing it in the same way as the quasi-3D velocity-field in Fig. 3.7, you can immediately see a difference in the flow pattern when compared to the other approaches. Now two eddies arise just after the channel has broaden. To understand this change, it is crucial to realize that all 2D models assume an infinite extension within the last direction. In this example, the flow would then broaden from an infinitely deep narrow slab out into a broader, but still infinitely deep, slab. Having this realistic interpretation of the 2D model, there would indeed be two rolls on each side of the transition between the two slabs.

Actually a calculation of the corresponding Reynolds number gives the value 10.4, which is in nice correspondence with the existence of the two rolls, and therefore not even the value of the Reynolds number will indicate that the pure 2D approach is an erroneous model of the given example.

What should prevent such an approach is to realize that it is always the smallest extension of fluid channels that dominate the flow properties of the channel, since viscosity can be interpreted as a diffusion of momentum from the channel walls into the fluid flow.

The impact of such a gross error can be illustrated by extending the "insight scale-bar" or "distance to realism" axis, as introduced earlier, into the "negative" direction, shown below in Fig. 3.10.

Now the bad numerical models increased the distance from "reality" beyond the initial concept, since we may be misled in the further investigations, if they are based on such false results.

The extension of the error is clearly illustrated in Fig. 3.11 by once again plotting the height averaged flow profile, as in Fig. 3.8, but now across the whole width of

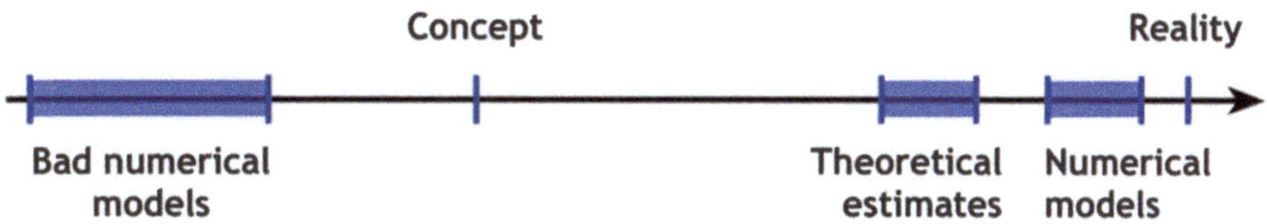

Fig. 3.10 Illustration of the impact of bad numerical models on the insight gained, by extending the "Distance to reality" axis, introduced in the beginning of the chapter

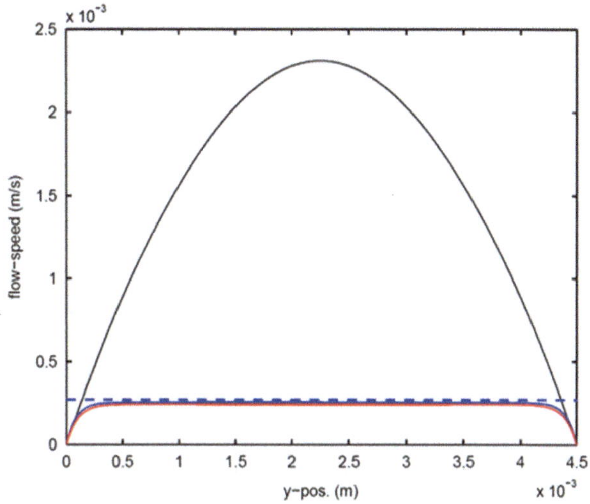

Fig. 3.11 The flow profiles gives by the different model. Compared to Fig. 3.8 the pure 2D result totally deviates from the other models

the broad channel and including the result from the pure 2D model as a solid black curve.

The pure 2D velocity profile strongly deviates from all the other approaches, and that nicely emphasizes the point that it is a good habit to do a quick check of the numerical results using for example a simple lumped element model (given by the dashed blue line in Fig. 3.11 that are barely distinguishable from the exact velocity profile) instead of wasting many hours or even days pursuing the implications based from erroneous results.

3.6 Modeling Convection and Diffusion

One of the most common modeling problems in microfluidics is the transport of species in a microfluidic channel. Typical problems include mixing of two concentrations or studies of hydrodynamic focusing. In all these cases the flow problem

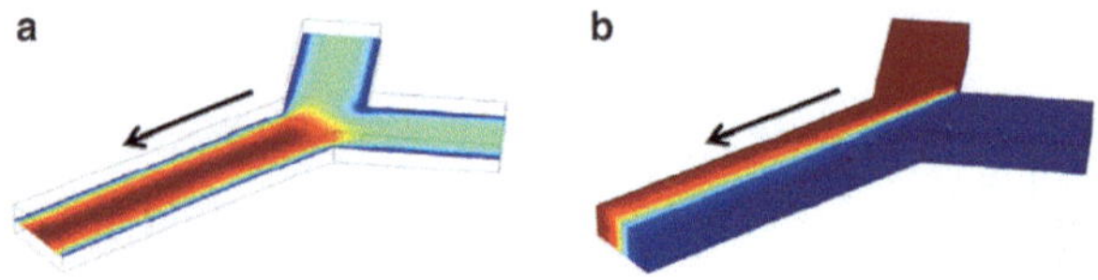

Fig. 3.12 (**a**) The obtained velocity profile for a two inlet system exiting in one outlet. The two inlets have identical hydraulic resistances. (**b**) The obtained concentration profile for this system, where one inlet has been set to concentration of 1 mol/m^3, while the other at 0. The result is in this case also constrained within these two limits, as expected

can be solved separately and the solution for the velocity can then be used as input for the concentration problem.

In these types of models users usually experience that the concentration they obtain after solving the problem sometimes is negative or much higher than the maximum concentration they had set as input. This is a very common numerical error based on the fact that the software cannot easily deal with abrupt changes in the variable under investigation.

Let us illustrate this with an example. Take the geometry of Fig. 3.2 and reverse the flow, so that it now is a two inlet system, with the two inlets merging into a single outlet channel. If we model the system like it is shown in Fig. 3.2 we will notice that the concentration is very nice and smooth with no overshooting at any point apart from a slight "unsmoothed" region at the corner where the two inlet channels meet. This is shown in Fig. 3.12.

Although modeling this two inlet system like it is shown in Fig. 3.12 is by no means difficult it can be of interest to skip the modeling of the forked inlets and just split the inlet boundary of the outlet channel into two, half with a concentration of 1 mol/m^3 and the other half with a concentration of 0. This is a particularly attractive solution in case the main channel has some complicated structures that will greatly increase both the number of mesh elements but also the complexity of the velocity profile.

In such a case you can define your concentration at the point (boundary) where the two inlet channels meet as shown in Fig. 3.13a. If we take a line profile of the concentration across the boundary it will look like a step function, as shown in Fig. 3.13b. This is usually very easily definable by using logical functions to specify the concentration, e.g., $c = 1$ when $y > 0$, for the case of Fig. 3.13a. However, such a function cannot be produced numerically and the software will try to fit it to a continuous function, as shown in Fig. 3.13b. Therefore your boundary will suddenly contain over- and undershooting of the concentration and the obtained result will not be bounded by 0 and 1 as you would expect.

The effect of this to the concentration profile obtained will of course be greatly dependent on the mesh you use on this boundary as well as the type of problem you use (stationary or time-dependent). A very fine mesh will eliminate this error to a great degree but will cost you in computation time. Time-dependent problems will also show this error even though their stationary counterparts do not—again,

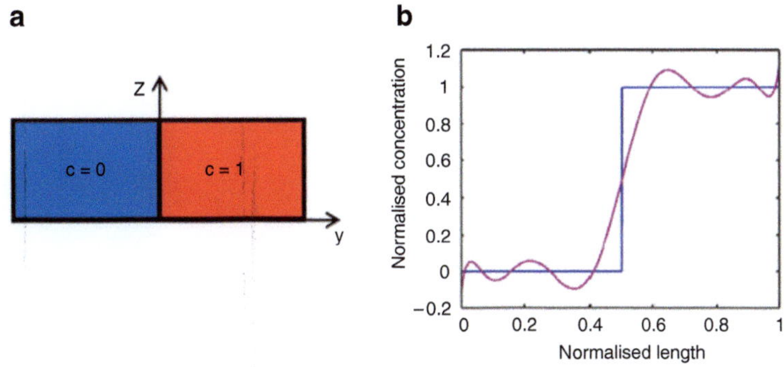

Fig. 3.13 (**a**) Schematic of a divided boundary for modeling two inlets with different concentrations. (**b**) Line profile of the actual concentration across the divided boundary (step function) along with the real concentration as will be fitted by the software in order to find a solution

dependent on the mesh size. The geometry of the problem will also play a role, as well as the software that you use.

It is generally considered good practice to use functions to define the concentration at a boundary. Taking COMSOL multiphysics as an example one can define step functions, pulse functions, Gaussian functions and use these as an input to the concentration boundary. These are smoothed to begin with (by a factor we can decide ourselves) and therefore no undesirable fit occurs. However, one should take care of that the initial value is not unnaturally smoothed out by the input function. Figure 3.14 shows an example of the concentration profile obtained in the channel when using the boundary condition shown in Fig. 3.13a. The over- and undershooting is about 8 %, whereas it drops to under 4 % when using a function to define the boundary concentration. Although the difference is not that big in this particular simple channel, it can be significantly higher for more complicated geometries, with an overshooting of more than 20 % when using logical input at the boundary instead of functions.

Simulation software is becoming better and many programs have worked around these issues and can therefore provide good solutions even though the boundary conditions are not optimal. The solution presented in Fig. 3.14 would look completely different only a handful of software updates ago. It is therefore far from certain that you will encounter such results in your simulations. However, should some overshooting arise in your solutions, it would be a good idea to check the mesh and the boundary condition definition. You should also consider that a really sharp boundary such as the one presented in Fig. 3.14 is not found in any device you may fabricate. A smoothed step function is therefore much closer to reality.

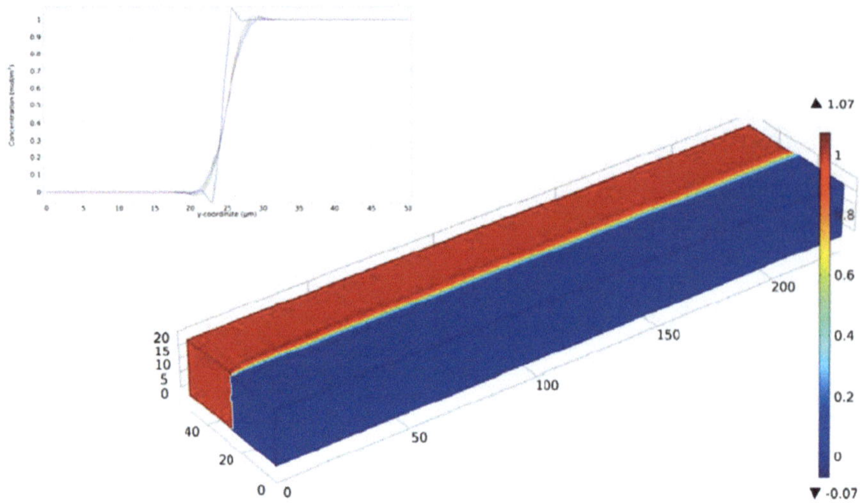

Fig. 3.14 Surface concentration in a splitting channel for a species with diffusion coefficient of 10^{-12} m^2/s, with the inflow concentration boundary set as a logical function. The inset shows the concentration profiles in lines along the x axis at a height of 20 μm. The overshooting is visible everywhere along the channel, but it is highest closest to the inlet. We also notice that the division line at the inlet is not sharp but follows the mesh elements. Therefore, your starting point may differ depending on the mesh size

3.7 The Importance of a Good Mesh

One of the most important parameters that can influence the solution of a certain problem is the quality of the mesh. In fact, this is also one of the steps in a simulation that may cause you the most problems. A bad mesh may not only result in a bad solution but also—more often—in no solution at all. Ironically, making a mesh is also one of the most difficult parts of a simulation, not only for beginners but also for more advanced users.

A very crude guideline is that the larger the number of elements, the better the mesh in terms of obtaining a correct solution. However, this only holds true to a certain degree. At some point in time further increasing the mesh size will only increase the computational complexity without really offering any advantages for the solution.

Let us illustrate this with a simple example. Assume that you have a microfluidic channel that is 20 μm high and 200 μm long, while the width is much larger than the height. We can therefore approximate the flow in a 2D geometry. The expected flow is of course laminar and therefore we expect to see a parabolic flow profile across the channel height. The input flow rate for a channel that is 50 μm wide is set to 1 ml/h. This corresponds to a 2D flow rate of 5.5556×10^{-6} m^2/s, which should be constant along the channel length. Figure 3.15 shows the calculated flow rate in units of % of inlet flow rate in the middle of the channel as a function of mesh size.

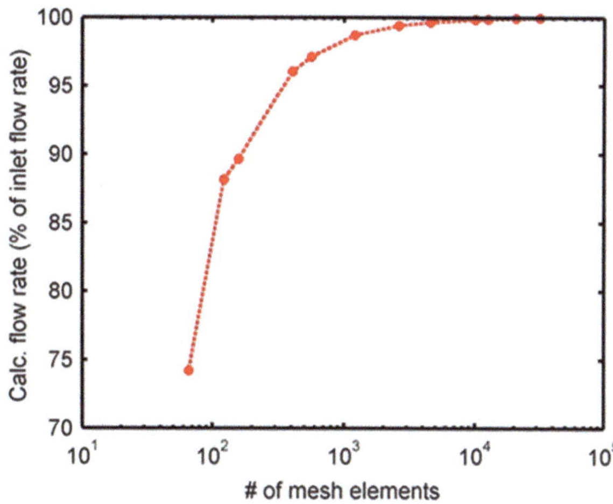

Fig. 3.15 Mesh convergence data for the calculated flow rate compared to the input flow rate. There is a great improvement by increasing the number of elements from 70 to 4,000

Clearly, we cannot get the correct flow rate by using only a few number of elements. Increasing the number of elements from 60 to about 4,000 provides a huge improvement to the solution. Further increasing the number of elements to 30,000 does not change the obtained solution significantly, so increasing the number of elements above 4,000 only contributes to a greater computational load and not to the improvement of the solution. This mesh convergence graph is something you should do at least once when you have a new problem. The parameter that you monitor is something that should have a well-known value that you can compare your solution to.

Another guideline for creating good meshes is to make sure that your mesh is detailed, i.e., use smaller mesh elements, where you expect your solution to change rapidly spatially. If you expect your solution to change from 0 to a maximum value for example within 2 μm from a certain boundary, then it will be a good idea to include at least a few elements in those 2 μm, otherwise you will not be able to resolve the change. Most software available lets you manipulate the mesh on every level, i.e., domain, boundary, edge, or even point, so such action is possible, as long as you have a certain idea about how your solution should look like.

Let us illustrate this with another example. Assume again that you have a microfluidic channel that is infinitely wide so that you can do the simulation in 2D. In a small part of the channel bottom and top you have a concentration (1 mol/m^3) of small molecules diffusing into the channel with a diffusion coefficient of 10^{-12} m^2/s. You are interested to see what the concentration of your species will be in the channel at steady state conditions given a certain flow of liquid from left to right (see Fig. 3.16).

Clearly, the most interesting part from the diffusion point of view is the region immediately above and below the bottom and top diffusion inlets. To obtain a good solution for the steady state concentration in the channel we should therefore have a thin mesh close to these two boundaries. For the sake of the argument we will create

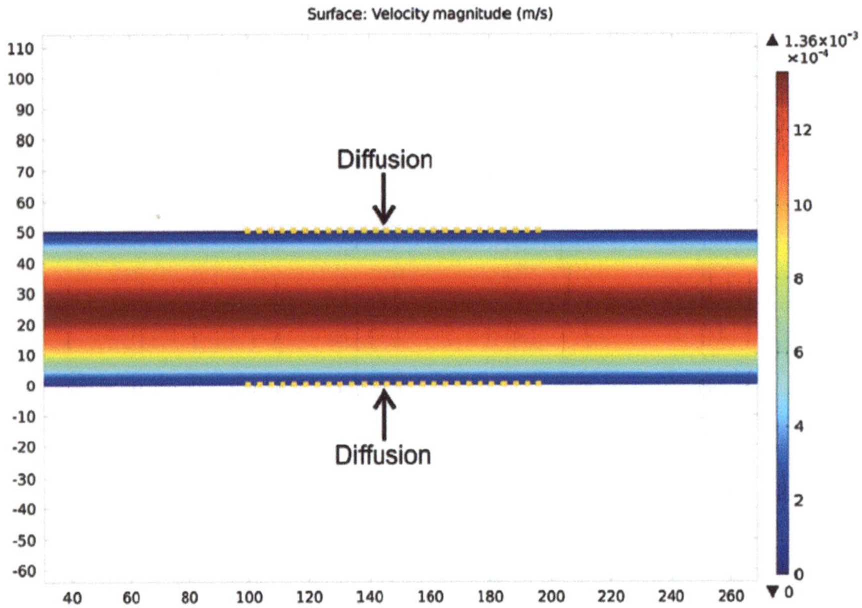

Fig. 3.16 The velocity profile in the microfluidic channel. The two diffusion inlets are marked with *dotted lines*. The axes are distances in μm. The inlet is at 0 μm (x axis) and the outlet at 300 μm (x-axis)

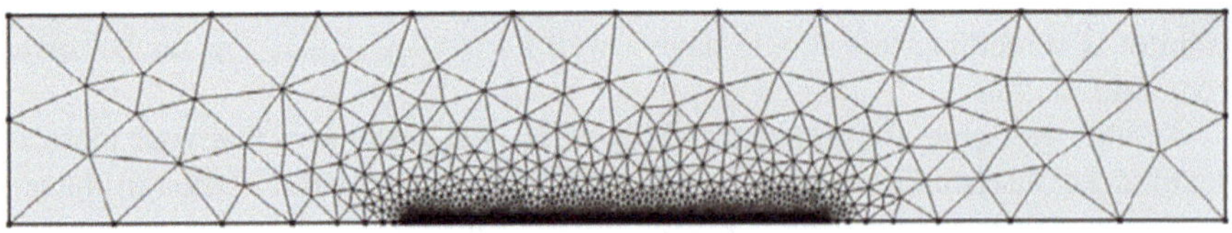

Fig. 3.17 The mesh of the domain for solving for the concentration. For the sake of the argument we have used a very fine mesh on only one of the diffusion boundaries

a thin mesh only on the bottom boundary and a very crude mesh at the top boundary, as shown in Fig. 3.17.

The solution of the problem using the mesh of Fig. 3.17 is shown in the inset of Fig. 3.18a. As the problem is completely symmetric, there should not be any differences in the concentration between the bottom half and the top half of the channel. This is clearly not the case and the diffusion from the top boundary which was poorly meshed is not only not well resolved but also results in the calculation of negative concentrations. Figure 3.18a shows the concentration profile at five different points in the channel. The same graphs are presented in Fig. 3.18b when the mesh on the top boundary is also thin. You can clearly see that choosing a mesh that can resolve the physics you are solving is very important in this case.

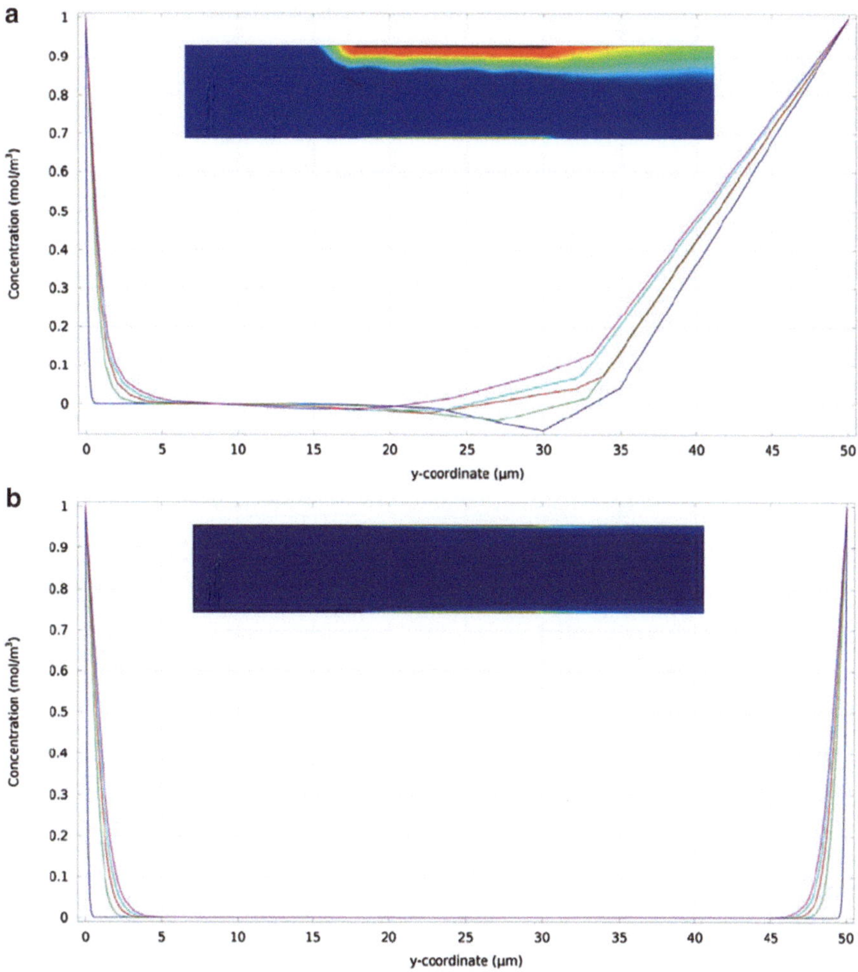

Fig. 3.18 (**a**) Solution of the convection/diffusion problem when using the mesh of Fig. 3.17. The effect of the mesh size can clearly be seen: although the concentration profile on the bottom boundary has been resolved, the same is not the case for the much rough meshed top boundary. (**b**) The correct solution of the problem, when using a good mesh on both boundaries

3.8 Software and Further Reading

As has been mentioned in the previous sections, there are a lot of different simulation programs available that can perform numerical calculations. Table 3.2 presents a few of these, although the list is not by any means complete. Which one you choose depends mainly on your simulation experience, knowledge of the program by you or your colleague as well as by license availability and price.

Table 3.2 Some of the available software for microfluidic simulations together with a brief description

COMSOL multiphysics	Very versatile software, used in every conceivable aspect of physics. Very intuitive, no need for deep theoretical knowledge to get started. Equation mode is also available if you prefer to control your own simulations. Live link to 3D drawing programs. Own geometry drawing module can be challenging. Many tutorial models available that can get you started. Solution times vary with available hardware and can be large for complicated simulations
Coventor	Program specifically oriented for simulations in microtechnology (MEMS and microfluidics). Best for MEMS. Can be a bit difficult to start with, but is otherwise very effective
ANSYS (fluent)	Part of the larger software ANSYS for fluid simulations (all types, incl. microfluidics). Interactive solver setup, possibility for pausing the calculation and change a parameter before continuing. Usually solves faster than COMSOL or ACE+
CFD-ACE+	Multiphysics simulation tool. Quite intuitive. Similar to COMSOL though with a bit more strict requirements on the correct setup of the numerical parameters. Can be slow for complicated models
Flow 3D	Powerful software specifically for flow simulations, incl. all types of interactions, e.g., electroosmosis, fluid–structure interaction. Very effective meshing, reducing computational times

Table 3.3 A few examples of publications utilizing numerical simulations

Publication	Details
(Suh and Kang 2010)	A review on mixing in microfluidics, showing theoretical and experimental results
(Boy et al. 2008)	More general paper dealing with the type of simulations that you can do along with a few examples
(Adam et al. 2012)	Article dealing mainly with mixers, showing both theoretical and experimental data
(Moosman et al. 2008)	Article comparing different software for simulations of the same structures
(Ramos et al. 2012)	Numerical simulations of fluid mixing by electroosmosis
(Dimaki and Boggild 2004)	Simulations of dielectrophoretic manipulation of carbon nanotubes

Some programs are not limited to microfluidic simulations but are more general, while others are very specific to fluids.

There is a lot of literature showing how simulations help the design of lab-on-a-chip systems or provide explanations for experimental results. Some examples are presented in Table 3.3.

3.9 Conclusion

In this chapter we have tried to give an introduction to how one should approach simulations of microfluidic systems. If you are a beginner in the field, this chapter has hopefully provided some insight into how to start your simulation and how to proceed, with the focus being primarily on the most common sources of errors and the suggested solutions. The most important take home message is that simulations *can* and *do* provide false results if they are not set up properly. The key to success lies in following these few steps:

1. Always start by taking a piece of paper and a pen and do some simple calculations that will give you insight into the nature of your problem. The simplest thing to start with is to calculate the relevant dimensionless numbers.
2. Check how you can simplify your geometry properly by utilizing symmetries, quasi-2D if applicable, lumped modeling.
3. Carefully consider your input parameters, boundary and domain settings.
4. Create a mesh that will provide you with the level of detail you expect.
5. Always verify that the solution you get also makes sense.

References

Adam T, Hashim U, Diyana PA (2012) Numerical simulation of microfluidic devices. J Appl Sci Res 8(4):2162–2174

Boy DA, Frederic G, Sumita P (2008) Simulation tools for lab on a chip research: advantages, challenges and thoughts for the future. Lab Chip 8(9):1424–1431

Dimaki M, Boggild P (2004) Dielectrophoresis of carbon nanotubes using microelectrodes: a numerical study. Nanotechnology 15(8):1095–1102

Moosman C, Glatzel T, Litterst C, Cupelli C, Lindemann T, Niekrawletz R, Streule W, Zengerle R, Koltay P (2008) Computational fluid dynamics (CFD) software tools for microfluidic applications – a case study. Comput Fluids 37(3):218–235

Ramos A, Loucaides N, Georghiou G (2012) Configurable AC electroosmotic pumping and mixing. Microelect Eng 90:47–50

Suh YK, Kang S (2010) A review in mixing in microfluidics. Micromachines 1(3):82–111

Chapter 4
A Considered Approach to Lab-on-a-Chip Fabrication

G.D. Kipling, S.J. Haswell, and N.J. Brown

Abstract This chapter details the common fabrication processes for lab-on-a-chip (LOC) devices. Particular attention has been paid to how the functionality of these devices can be enhanced though the adoption of a considered engineering based approach to chip fabrication which encompasses the design, material selection and manufacture process stages. Fabrication considerations in this chapter extend to the potential for mass production of LOC devices using a range of materials and manufacturing methods. The adoption of more time and cost-effective fabrication methodologies is critical to the future integration of these devices into mainstream applications. Additional factors such as device repeatability, dimensional tolerance and quality control, all of which vary between fabrication processes, are also considered and discussed.

4.1 Introduction

Lab-on-a-chip devices are increasingly being developed globally by research groups and commercial ventures that require the performance of such devices to become ever more sophisticated. Accordingly, it becomes more difficult to provide a 'one-size-fits-all' device that can be used across the diverse interests that range from biomedical to environmental and industrial applications. Whilst it may be possible to realize simple research objectives on a generic chip, the development of integrated or enhanced functionally devices will increase the requirement for a bespoke, engineered approach to the fabrication process. As with any engineered components, there are a multitude of different fabrication possibilities ranging from material selection and design through to fabrication methods and machining parameters. This wide range of engineering approaches coupled with the diversity of applications for lab-on-a-chip devices can make the process of fabrication rather daunting. In addition, the academic and industrial fields that benefit most from the utilization of lab-on-a-chip devices are, amongst others, chemistry, biology and

G.D. Kipling(✉) • S.J. Haswell • N.J. Brown
Universityof Hull, Cottinham Road, Hull HU6 7RX, UK
e-mail: garydavidkipling@gmail.com

J. Castillo-León, W.E. Svendsen (eds.), *Lab-on-a-Chip Devices and Micro-Total Analysis Systems*, DOI 10.1007/978-3-319-08687-3_4

environmental sciences, all of which may not have a particularly strong understanding or appreciation for engineering processes. It is for this reason that engineers are becoming more active in lab-on-a-chip research and why an interdisciplinary approach to chip design and fabrication is essential. The integration of engineering knowledge from the outset of a project will not only offer the potential to enhance the functionally of the chip as well as identify opportunities for expansion and further improvement but will also ensure the outcome of such research will lead to the viable manufacture and thus, the commercial success of a device.

From woods to metals to polymers it is clear that most materials have their place in society, which may be due to their relative low cost, high strength or ease of manufacture. The same is true for the lab-on-a-chip device. Whilst it may be possible to immediately disregard some materials, there are no generic 'correct' materials that should be utilised in their fabrication. Each material has its relative advantages and disadvantages, all of which must be considered as part of the design and fabrication process to ensure the device is successful in fulfilling its functional criteria. It may be possible to fabricate a chip from a number of different or combined materials or through different manufacture processes, increasing the importance of a considered engineering based approach. In addition to the physical properties of a finished device, the material selection process for a chip must consider the possible manufacture methods that are applicable. When undertaking small scale production of chips, this consideration may only extend to the capability of a department or company, however, when large scale fabrication is required, this factor is likely to carry particular significance during the design process. Failure to fully address these issues could have severe consequences with regards to production rates, unit cost and future device modifications.

This chapter will show that the key to successful chip fabrication is a considered approach throughout all the stages of design and manufacture, in addition to an understanding and appreciation of the likely compromises that will be required as a result of any decisions made during the selection process.

4.2 Engineering Design

Whilst this chapter is dedicated to the fabrication of lab-on-a-chip devices, it is important to stress the significance of engineering design as a means of ensuring efficient and effective manufacture of chips. As such, this brief section will discuss some engineering design considerations that are vital in ensuring a successful fabrication process.

It is clear from the literature that the ability to perform laboratory operations on a lab-on-a-chip device is the key to its appeal (Whitesides 2006). As such, the design of any lab-on-a-chip device must, first and foremost, consider the functional requirements from a chemical or biological process standpoint. Consider for example, a fundamental physical feature present in a majority of devices; the channel.

Geometries such as length and width may be dictated by the requirement for diffusive mixing of two reagents. Failure to provide suitable channel dimensions could potentially inhibit the performance and thus limit the success of a lab-on-a-chip device. Whilst the physical features on a chip are designed to enable or enhance the functionality of processes, their fabrication is essentially though the application of engineering methods.

Engineering considerations are meaningless if the functional requirements are not achieved, however, their inclusion in the design process has the potential to further augment the usefulness of a lab-on-a-chip. These benefits range from a simplified or more time efficient manufacture processes through to the capability of fabricating complex, integrated devices with multiple functionalities. As such, it is unlikely that the full potential of lab-on-chip device will be realized without the consideration of both process and engineering factors at the design stage.

Recognizing the relationship between functional requirements of a device and the engineering design at the conceptual stage allows for the identification of potential problems relating to the manufacture and utilization of lab-on-a-chip devices and as such is key to the future development of high performance lab-on-a-chip technology. The following sections are dedicated to the manufacturing material, process and assembly methods that are commonly used in the fabrication of lab-on-a-chip devices. Aspiring chip fabricators may be able to consider and evaluate these manufacture processes in isolation to produce a functional device. However, it is the integration of these considerations during the design stage that is vital for the optimization of functionality and fabrication capability. These considerations extend to the possible geometric accuracy that is achievable using the various methods. An important theme highlighted throughout this chapter will be how the allowable tolerance (i.e. the acceptable accuracies associated with the fabrication) for a particular device may have a substantial influence on which material, manufacture process or assembly method is suitable for fabrication.

4.3 Manufacture

Fabrication of lab-on-a-chip devices could simply be defined as the physical production of devices or products; however, as indicated it is important to consider the fabrication process as being inclusive of the engineered design and material selection stages. Despite this, the term fabrication will inevitably be primarily associated with the manufacture and assembly of the devices prior to use or sale. It may also be necessary to consider the fabrication process of 'on-chip' features (pumps, values, sensors, etc.) which will have an influence on the functional requirements of the device. It has already been stated that there is an abundance of possibilities when contemplating the fabrication of chips, in particular with regards to the manufacturing processes. It is important to recognize the compatibility of materials with particular manufacturing processes and the considerations that are required to ensure that the functional requirements of the devices are

Table 4.1 Manufacture and assembly methods for a range of applicable materials

Material	Manufacture method	Assembly method
Glass (Iliescu et al. 2012)	Photolithography	Adhesive
	Chemical etching	Glass fusion
	Micromachining	
	Powder blasting	
	Water jet cutting	
Rigid polymers (Becker and Locascio 2002)	Micromachining	Adhesive
		Thermal bonding
	Injection moulding	Ultrasonic welding
	Hot embossing	
	3D printing	
Soft polymers (McDonald et al. 2000)	Cure moulding	Adhesive
	Soft lithography	
	Rapid prototyping	Contact bonding
Paper (Yetisen et al. 2013)	Wax patterning	Adhesive
	Alkyl ketene dimer (AKD) printing	
	Flexographic printing	
	Shaping/cutting	

achieved. The primary materials discussed in this section and their relevant methods of manufacture and assembly are shown in Table 4.1. It should be noted that in many instances it is possible, perhaps necessary, to utilize two or more of the manufacture or assembly methods for any given material.

4.3.1 Glass

It is perhaps unsurprising, given the history of glassware in chemistry laboratories, that glass has been adopted as a key material in lab-on-a-chip fabrication in the field of chemistry. Optically transparent and with excellent chemical resistance, glass has the additional benefit of being familiar to the traditional 'off-chip' chemist, making the transition to microfluidics easier. Experiments utilizing solvents, acidic or alkali reagents will typically be more compatible with a glass chip than, say, one fabricated from a polymer. In addition there is a considerable history of surface modification and functionalization associated with utilizing glass as a substrate. Whilst this summary is somewhat broad and simplistic, the fact remains that if reactions can successfully take place in laboratory glassware, they are also likely to be compatible with a glass chip, thus somewhat negating the question of chemical compatibility in the design and fabrication process. Of course, the question of compatibility on glass chips cannot be completely disregarded since factors such

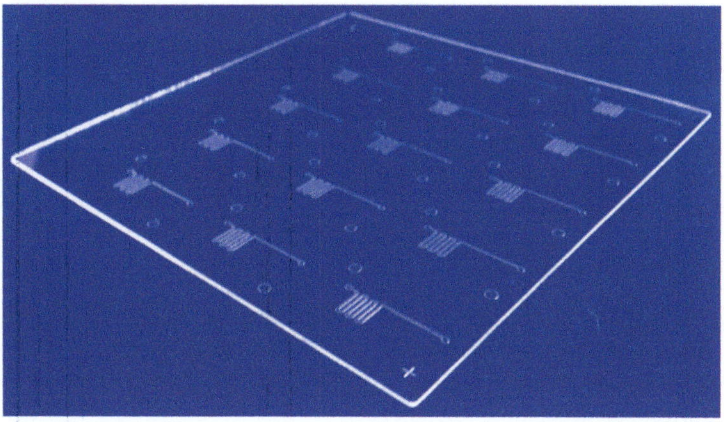

Fig. 4.1 Example of a glass wafer with wet etched channels

thermal considerations are not necessarily comparable when applied on a micro scale (Fig. 4.1).

Although processes used in glass fabrication, such as photolithography and chemical etching, are well established, they rank relatively low when taking into account mass production and fabrication efficiency. It is for this reason, amongst others, that glass is generally viewed as a specialist engineering material and one that is largely incompatible, with traditional machining process (turning, milling, etc.). Undertaking a structured material selection process from an engineering standpoint would typically favour 'traditional' production methods such as injection moulding, forming or stamping, all of which are usually not suitable for glass. As a result of this challenge to utilize mass production methods, the financial cost of a glass chip is likely to be considerably greater than some alternatives (e.g. injection moulding of polymer devices). However, as previously stated, the inability to mass produce large quantities of chips may not be a concern for certain applications; for example, in some research environments. The utilization of glass in the fabrication of commercial products could potentially be inefficient and costly relative to alternative materials. Conversely, it may be easier and potentially more cost-effective to fabricate a small number of microfluidic prototype devices in glass rather than, for example, having to design and manufacture mould tools for polymer device fabrication at each stage of chip development.

Key relative advantages and disadvantages associated with the use of glass in chip fabrication are listed below.

Advantages

- Chemical and thermal compatibility enable a wide range of functions to be undertaken on the device, including those that require aggressive solvents, corrosive reagents, surface modification or high temperatures.
- Optical transparency is very good allowing for analytical functions such as chemiluminescence to be performed on chip.

- Glass is compatible with electrophoretic based separation applications.
- Physically, glass is rigid which has the potential to better maintain the integrity of channels and features on the device.
- Assembly methods such as glass fusion are preferable for high pressure applications due to their relative high strength.
- Biocompatibility of glass is desirable for application that utilize biological processes or reagents.

Disadvantages

- Relative inability to utilize effective mass production methods limits the speed and efficiency with which the device can be fabricated.
- Specialist machines and manufacture methods are required to ensure devices are made within an acceptable tolerance.
- Relative high unit cost associated with raw material and applicable manufacture methods.
- Fragility of the material under flexural or tensile loading increases the risk of cracking or fracturing the device. This is of particular concern for medical applications where failure of the device could spread biological material.
- Chipping of the material is a concern during manufacture, in particular around holes. This can also lead to inherent structural weaknesses in the device.
- Bonding glass through thermal fusion is slow and requires very high temperatures.

Photolithography and Etching

Utilizing photolithography in combination with wet etching in glass chip fabrication has the potential to afford the user a great deal of flexibility with regards to the channel or feature designs that can be attained. The first stage of this process involves the production of a photo mask that is used to transfer patterns to a photo resist coated glass wafer which then has channels and features wet etched into the surface to a required depth. It should be noted that the etch depth achieved using this approach will be uniform across the chip but the width of the channels or features will vary depending on the mask design. Chip features, such as channels are typically designed using computer-aided design (CAD) software such as AutoCAD (Autodesk, Inc., USA). Adoption of CAD software by chip designers has facilitated an increase in the accuracy of chip features and channel geometries, further enhancing the potential for lab-on-a-chip device fabrication to be repeatable. Devices designed using CAD software can also be easily scaled to produce chips of different bulk geometries whilst maintaining the same feature and channel proportions. Ensuring accurate physical geometries of channels and features on a device also helps to increase confidence in any prior simulation results regarding fluid flow or chemical interactions.

Glass wafers are typically coated with an adhesion layer (e.g. chromium or a commercially available primer) prior to the application of the photoresist to ensure

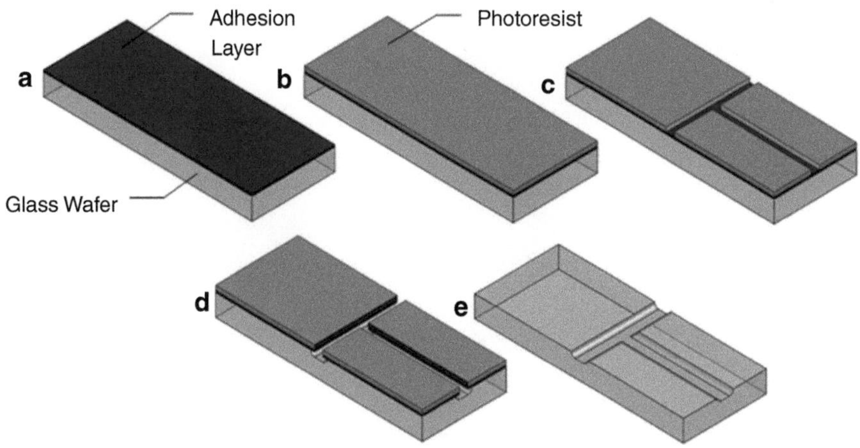

Fig. 4.2 Schematic representations of the wet etch procedure using an isotropic etchant

that it is not delaminate during the etch process (Fig. 4.2a). The chromium also reduces etch undercut and improves the quality of the channels produced. A photosensitive material, typically a polymer, is then applied to the coated substrate through a process such as spin coating (Fig. 4.2b). Selective exposure of the photoresist, for example through a mask of the predesigned chip, to UV light increases the solubility of the material which can then be dissolved when it is come in contact with a solvent (Fig. 4.2c). Immersion of the glass wafer in a corrosive medium (e.g. hydrofluoric acid) (Iliescu et al. 2008) etches into the glass wafer, thus creating channels in the chip (Fig. 4.2d). Finally, the photoresist is removed and the adhesion layer is stripped using an appropriate etchant (Fig. 4.2e) (Stjernstrom and Roeraade 1998).

When designing the mask it is important to consider how the etchant will remove material since the cross-sectional profile of an etched channel is liable to vary depending on the material and etching method. It is not uncommon for an etchant to remove material in all directions (i.e. width and depth of the channel) thus creating an isotropic etch profile. It is clear from considering the fundamental physical results of using such an etchant, that the desired channels from the mask will not be simply transposed onto the material. A perfectly isotropic etch will remove material equally in all directions, thus resulting in a channel cross-section that has a low aspect ratio as demonstrated in Fig. 4.3. Channel width (W_C) can be estimated from the mask width (W_M) and the channel depth (D_C) (typically estimated as a function of time through experimental extrapolation) which are related by the function:

$$W_C = 2D_C + W_M \tag{4.1}$$

This relationship can be expanded to equate the achieved geometrical dimensions with time if the etch rate is known. Consideration of this relationship at the design

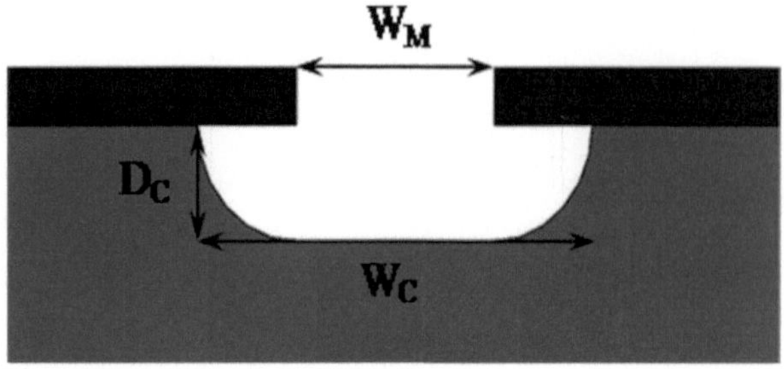

Fig. 4.3 Schematic cross section of a channel fabrication using an isotropic etch detailing the approximate relative dimensions

stage is crucial to ensure that the etched geometries are of the correct width and depth. It is also clear that channels fabrication using isotropic etchants will inherently have rounded corners at the base of the channel or feature with a radius roughly equal to the channel depth. Consequently, it is not possible to fabricate channels with sharp corners through the application of this method. Poor surface quality of channels or features produced through wet etching can lead to tortuous fluid paths, especially when relatively wide channels are required. Simple fluid delivery may not be adversely affected by poor quality features; however, this could affect more complex functions such as diffusive mixing or the requirement for a specific flow rate.

Alternative etching profiles such as anisotropic or straight-walled can be achieved through the use of different fabrication methods such as reactive ion etching (RIE) or through the use of different etchants. These methods are primarily applied to silicon rather than glass wafers due to the relative high cost and slow production rate when compared with wet chemical etching.

Regardless of the etching method used or the attained channel profile, it may be necessary to consider integration of the etched wafer with additional fabrication methods. This may be as simple as ensuring there is enough space between individual chips to allow for manual sectioning, or a considered and perhaps intricate arrangement of alignment features that will enable further processes to be performed on the wafer. Such processes could include the addition of further features with computationally determined geometries that have a very small tolerance. Alternatively, the wafer may be sectioning using precision machining techniques (discussed in Sect. 4.3.1.3) which require an alignment position relative to the etched channels. Additionally, it is necessary to consider the space between features to ensure an adequate surface area is available for bonding during the chip assembly process (Abgrall and Gue 2007; Chen et al. 2007; Iliescu et al. 2008; 2012; Stjernstrom and Roeraade 1998).

Micromilling Glass

Considerations associated with micromilling components or devices vary with the material being used for their construction. Metals, which have a long history of being processed through milling methods, typically require a scaled down approach to a well-established machining procedure. Parameters such as the tool spindle speed and feed rate must be changed and optimized depending on the feature size and raw material; however, this is relatively straightforward to the experienced machinist.

As previously discussed, glass can be considered a specialist engineering material due to its relative incompatibility with standard manufacture methods, including milling. Thus, the effective fabrication of glass lab-on-a-chip devices through micromilling requires more attention than other materials. Standard fluted milling tools are not suitable for glass milling, primarily due to the relative hardness and low tensile strength of the material. In order to prevent or at least limit chipping and cracking of the work piece, diamond coated tools are used as an alternative. These tools have a thin coating of industrial diamonds that effectively act as an abrasive surface when the tool is spinning at high speeds. The relative small scale of the diamonds combined with their inherent hardness aids in applying the milling process to glass. Figure 4.4 shows a scanning electron microscopy image of a 2 mm diameter diamond coated milling tool detailing the individual diamond burs. Feed rates and spindle speeds are the main variables that require optimization for each grade of glass and feature design. Failure to thoroughly address these parameters can lead to a number of issues relating to feature precision and loss of quality including:

- Hole chipping, which can occur at the point where the tool first contacts the work piece (break-in chipping) or on the underside of the material (break-out chipping). Chipping can lead to an increase in effective hole diameter which has the potential to lead to problems with connection compatibility.
- Surface roughness as a result of milling is inevitable when using a diamond coated tool, however, it is important to minimize this were possible. Excessive

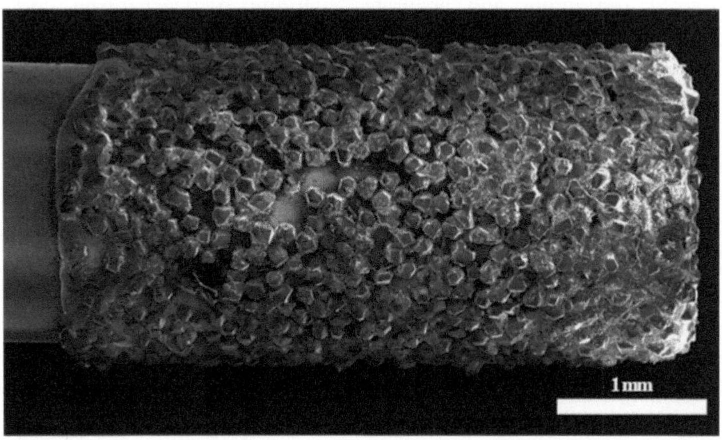

Fig. 4.4 Example 2 mm diameter diamond coated end mill suitable for micromilling of glass wafers captured using a SEM

roughness of features can lead to issues relating to the flow and overall functionality of a device.
- Cracking of the work piece can occur as a result of excessive force being applied to the glass during the milling process. Even if the cracks are oriented so as to not affect the functionality of the device it may still be rejected due to the inherent weakness caused as a result of their presence. Further cracks may propagate from the weakness as a result of applied pressures or fatigue effects.

Any of the aforementioned issues can be exacerbated by tool wear. Ensuring the correct machining parameters are established and adhered to is also an effective method of prolonging tool life and thereby further reducing the number of devices that are rejected during quality control checks. Overheating during machining is a particular concern with glass milling since this can cause very rapid degradation of the tool through stripping of the diamond burs from the metal substrate. Tool damage of this nature is likely to lead to cracking of the work piece as a result of an increase in applied load. The application of an effective cooling method, for example ethanol misting or water flooding, is therefore necessary in addition to the optimized machine parameters.

Most micromilling is performed using a computer numerical control (CNC) machine due to the intricate nature of the features being produced. Utilization of such a machine typically presents a great deal of flexibility with regards to the creation of features in three-dimensions. Micromilling allows for the production of relatively deep and thus high aspect ratio features, especially when compared with a process such as wet etching. The milling depth of a tool is typically quoted as being three times the diameter of the tool; for example, a 500 μm diameter tool could be used to produce a feature that is 1,500 μm deep. Additionally, the application of milling techniques allows for a variety of depths to be machined, which could extend to the creation of access holes and microfluidic channels on a single wafer using the same machining program. CNC machining has the additional benefit of being highly repeatable and accurate, enabling devices to be consistently manufactured to a typical accuracy of ± 10 μm with current, commonly available CNC equipment. Positional accuracy of a CNC machine is critical for ensuring the device features are the correct geometry and in the desired location. Whilst the positional accuracy of the machine may be ± 10 μm, it is important to acknowledge that the achieved feature precision may vary by an order of magnitude (± 100 μm or greater) due to machining flaws such as chipping and microcracking of the work piece. This potential reduction in achieved precision stems from the inherently limited machining resolution of glass using this manufacture process.

Precise machining also has the potential to enable compatibility with other fabrication techniques. For example, consider micromilling of access holes in a pre-etched wafer. Positioning the holes in the correct location on the device is critical in ensuring the flow of reagents is achievable and not restricted. Depending on the size and quantity of channels and holes, the geometric tolerance afforded to the machinist may be very small, in some cases less than 100 μm in the case of glass. Achieving such tolerances using manual drilling methods may not be possible, thereby further emphasising the advantages of CNC machining (Sein et al. 2004; Yan et al. 2002; Zhou et al. 2006; Zong et al. 2010).

Wafer Sectioning

Whilst it is physically possible to section glass wafers using a CNC machine, the slow feed rate combined with the relatively long tool path makes this largely impractical, in particular when taking into account fabrication efficiencies. It is therefore usually necessary to apply alternative manufacturing techniques when sectioning a wafer, either before or after CNC milling of features. Sectioning could simply entail traditional glass cutting techniques (i.e. scoring and snapping) or more advanced precision machining alternatives. One such applicable method for glass sectioning is through the use of a mechanical dicing saw. These machines utilize very thin (200–300 μm) diamond cutting blades and are capable of very high spindle speeds (30,000–40,000 RPM is typically suitable for glass). Positional tolerance of these machines can be as little ±10 μm or less making them suitable for high precision applications. Utilizing very thin blades with very fine diamond 'teeth' ensures that the potential quality of the sectioned edges can be very high; however, optimization is required for different wafer parameters (glass hardness, thickness, etc.). The use of a dicing saw or equivalent sectioning methods also ensures that the fabricated devices are repeatable and eliminates any potential for human error.

Alternative Methods

There are a number of less common methods that can be utilized in the fabrication of glass chips. Two alternative methods for the creation of holes in lab-on-a-chip devices are discussed below:

- *Powder blasting* involves the projection of fine abrasive particles (e.g. 10–30 μm alumina particles) at a high velocity on to the surface of the glass wafer. A mask containing the desired features (usually access/outlet holes) is applied to the surface of the work piece prior to blasting to prevent damage to the surrounding area. Features tend to narrow with increasing depth as a result of preferential abrasion nearer the wafer surface. Successful fabrication using this method is highly dependent on the abrasive particles used, pressure and work piece material; however, typically it is difficult to attain a high quality and precise features due to the relative hardness and brittleness of glass. Whilst an effective process for creating features or holes, the relative inflexibility of the method combined with the poor feature quality in glass make this technique less common than micromilling (Ghobeity et al. 2009; Park et al. 2004).
- *Water jet cutting* is an effective method of sectioning or creating through features (such as holes) in glass. Similarly to powder blasting, water cutting

involves the projection of water combed with an abrasive medium at high pressures. Fabrication of devices through this method is relatively uncommon due to its inability to create channels or other features and the high associated costs. However, it is a potentially effective method and viable alternative to mechanical dicing for wafer sectioning (Khan and Haque 2007; Nouraei et al. 2014).

Integrating Fabrication Methods

Previously, it has been indicated that it could be necessary or desirable to utilize numerous manufacturing methods on a single device. Indeed, the fabrication of some lab-on-a-chip device will utilize etching, CNC milling and dicing saw sectioning. Considered use of these techniques combined with optimized machining parameters has the potential to produce chips with very high geometric precision in the order of ± 10 μm. To take full advantage of this, it is important to develop a systematic and accurate methodology for aligning the work piece during each stage of the fabrication.

Alignment of micromachining tools is commonly achieved through either physical probing or visual positioning using inbuilt, calibrated cameras. Probing a 'raw' glass wafer (i.e. prior to any material preparation) can be problematic due to the manufacturing process by which the glass is produced. Dimensional tolerances of a glass wafer will typically be an order of magnitude greater than those achieve though precision micromachining. Additionally, ground edges, for example on the corner of the wafer, are not ideal sites for probing due to their relatively inconsistent geometry. There are two potential solutions to achieving a more accurate and consistent positional alignment:

- Initial sectioning of wafer, for example, with a dicing saw, prior to the application of any other fabrication process. This ensures that wafers have consistent dimensions as well as providing a 'sharp' corner which can be probed later in the fabrication process.
- Alignment features, for example those on an etched mask, allow for relative positions to be identified using visual alignment methods. This method, briefly discussed earlier in this chapter (Sect. 4.3.1.1), enables wafers to be sectioned and features milled in locations relative to the pre-etched channels/features.

Both approaches require additional processes to be implemented at different stages of the fabrication process and as such, the most suitable or efficient methodology will depend on the device being produced. Incorporation of such steps into a production process potentially leads to a decrease in production rates and increase in unit cost; however, this may be a necessary compromise if the device specifications require such exacting precision.

4.3.2 *Rigid Polymers*

For the purposes of this chapter, rigid polymers are defined as those that exhibit a sufficiently high young's modulus to provide a stable medium into which features may by moulded or otherwise imprinted. Further integration of lab-on-a-chip technology within both industrial and research sectors will likely serve to increase demand for these devices. It is therefore important to consider factors and limitation relating to the mass production of chips during the material selection process. Considering the relative strengths and weaknesses of possible materials quickly establishes rigid polymers as amongst the most suitable for mass production. The primary production method used in mass production of polymers is injection moulding (discussed in Sect. 4.3.2.2) which has the potential to produce large quantities of devices consistently, relatively quickly, cheaply and with little user input. Alternative methods for fabrication of polymer devices such as hot embossing are also possible, further increasing the number of potential applications that can take advantage of this material (Fig. 4.5).

Polymers have a wide range of different properties that can make them either highly desirable or entirely unsuitable for a particular application, for example with regards to the limited chemical compatibility of some materials. Common polymers that are suitable for use in the fabrication of lab-on-a-chip devices include but are not limited to; poly(methyl methacrylate) (PMMA), cyclic olefin copolymer (COC), polycarbonate (PC), polyethylene terephthalate glycol (PET), polyethylene (PE), polyimide (PI) and styrene copolymer (Becker and Locascio 2002).

A summary of the key advantages and disadvantages associated with the fabrication and application of rigid polymer chips is listed below. It should be noted that this is somewhat generalized and not all the points are applicable to all polymeric materials.

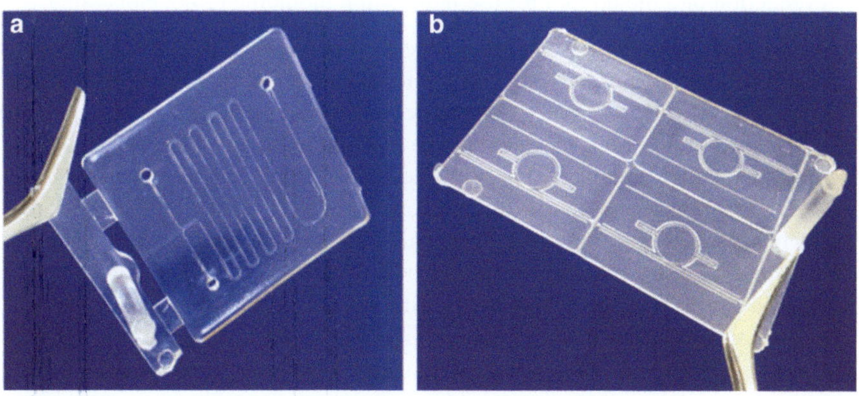

Fig. 4.5 Examples of injection moulded chips in (**a**) PMMA and (**b**) COC

Advantages

- Potential to produce large volumes of devices quickly, consistently, cheaply and efficiently using injection moulding.
- Moulding flexibility allows for the creation of complex geometries.
- Range of fabrication methods and materials allow for increased flexibility with regards to device design and application functionality.
- Optical transparency of some polymers enhances the potential range of analytical functions that the device can perform.
- Relatively inexpensive raw material. This is especially relevant with regards to injection moulding 'granules' which can be stored easily and used to produce a wide variety of geometries.

Disadvantages

- Some polymers demonstrate limited compatibility with solvents or acidic/alkali reagents.
- Specialist machines and expertise are required to take full advantage of increased production rates associated with polymer chips.
- Thermal resistance is relatively low compared to materials such as glass.
- Knowledge of specific material properties and manufacture parameters is required due to the extensive range of polymeric materials.
- Hydrophobic nature of polymers can potentially inhibit the performance of some devices as well as cause difficulties with surface functionalization.

Micromilling

Application of micromilling techniques to the fabrication of polymer lab-on-a-chip devices can be through two different methods:

- Direct removal of material from a stock polymer to form channels and features.
- Production of a mould tool for use in injection moulding or hot embossing.

As a consequence of the relative softness of polymers, direct micromilling is typically straightforward, providing correct machining parameters are established. For example, adequate cooling for the work piece is required at all times to prevent the polymer melting and causing excessive material build up on the tool leading to wear and potential breakages. It is not uncommon for polymers to have residual stresses within the material structure as a result of their composition and process with which they are manufactured. Removal of material through milling can lead to a redistribution of the internal stresses in the polymer piece leading to a warped or misshapen work piece and potentially altering the dimensions of any devices fabricated using this method. These stresses, which vary between polymer type and stock geometry, can be removed through annealing; however, this further complicates the fabrication process and is therefore undesirable when considering mass production of devices. Considering a primary benefit of utilizing polymeric materials in fabrication of chip is the mass production capability, it is therefore not surprising that direct micromilling of polymer stock material is relatively uncommon compared to alternative fabrication methods. Despite the relatively disadvantages of this method, prototyping of devices through direct micromilling of polymers can be time- and cost-effective, in particular when numerous different designs or configurations are required.

Typically, mould tools for either injection moulding or hot embossing are manufactured from a metal (e.g. steel or aluminium), which is an extension of a well-established manufacturing method. The use of very small tools (i.e. less than 1,000 μm) enables very intricate features to be moulding using these methods; however, it is important to ensure the associated machine speeds and feeds are adjusted to account for the relative fragility of the tools being used. Generally this will include an increase in spindle speeds as well as a reduction in the feed rate and step-down parameters. As with most other milling operations undertaken on this scale, moulds are usually fabricated on a CNC machine once a design encompassing all the necessary features has been generated. Software such as SolidWorks (Dassault Systèmes, France) can be used to draw a 3D computational model of the mould tool prior to manufacture. Integration with a computer aided manufacture (CAM) package such as SolidCAM (Dassault Systèmes, France) allows this drawing to be converted into a computation code (commonly referred to as a G-code) that can be recognized by the software on a CNC machine.

As previously stated, the typical machining depth achieved using the milling process is approximately three times the tool diameter; however, it is possible to achieve greater aspect rations with specialist fluted tools, albeit with a compromise relating to the maximum possible machining rate. It is therefore possible to manufacture devices with features or channels at different depths (Becker and Locascio 2002).

Injection Moulding

Children's toys, stationary, laboratory fittings and connections all take advantage of injection moulding to repeatedly produce high volumes of components or products at low cost. As discussed in Sect. 4.3.2.1, the first stage in producing a device using injection moulding is the creation of a mould tool. Manufacturing care and precision at this stage is critical, since any flaw or surface roughness of the mould will be transposed on to each device it is used to fabricate. The finished mould tool is inserted into the injection moulded before being clamped in place. Polymer granules are then fed from the hopper into the barrel and transported using the screw whilst simultaneously being heated. The softened polymer is then forced into the mould tool before being cooled and subsequently ejected from the machine. Figure 4.6 shows a simplified schematic example of an injection moulding machine highlighting the key components.

Fabrication of chips through the application of injection moulding also benefit from a high degree of flexibility, in particular with regards to the range of scales that can be incorporated into a single device. Small features, in some cases sub-micron,

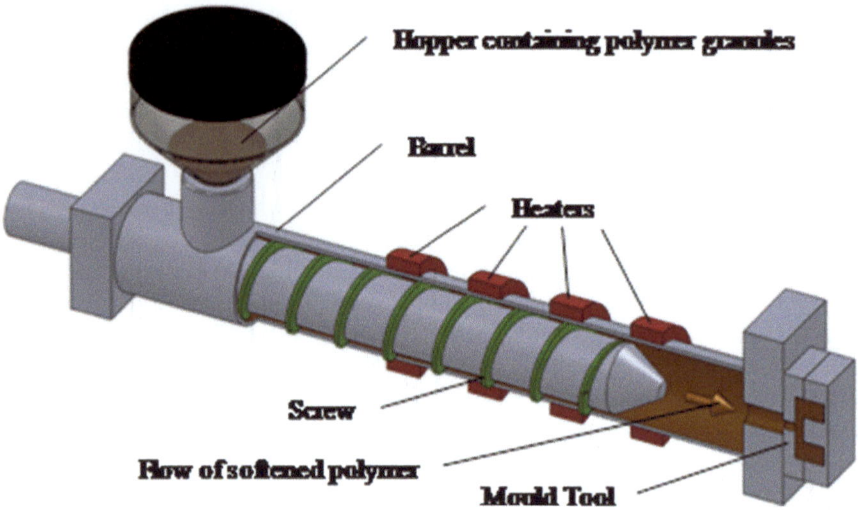

Fig. 4.6 Schematic cutaway example of the key components in an injection moulding machine

are possible due to the high resolution that can be achieved through machining of the metal mould. In addition, large features, such as reagent entry and waste holes can be moulded directly into the chip. Inclusion of holes and channels in the same mould has the advantage of negating the alignment concerns when drilling holes using a separate process or an alternative covering plate. Additionally, features with high aspect ratios can be attained from injection moulding, further enhancing the capability of this production method. Over moulding enables alternative features or materials to be integrated into the devices during the manufacture process. This negates alignment concerns when assembling the chip and potentially increases the functionality of the device. It should be noted that geometries machined into the mould are inverted when the polymer devices are ejected from the mould as shown in Fig. 4.7.

Exact replication of devices through this fabrication method can be complicated somewhat by a plethora of moulding parameters such as polymer or mould temperature, injection pressure or speed, shot size, cooling time, etc. Each parameter can affect the quality of the moulded device and as such must be optimized by the user and remain consistent throughout batch production. Application of inadequate parameters for a given mould can give rise to a large number of defects within the device such as sink marks, delamination, streaking, splay marks, warping, blistering or voids. Each defect is usually a result of a different moulding parameter which, once identified, should be altered in order to enhance the quality of the output devices. Material condition must also remain consistent, since contaminates such as moisture can have a negative impact (usually delamination of the material) on device quality. It may be necessary to dry some polymers in specific conditions prior to their use as production material (Fig. 4.8).

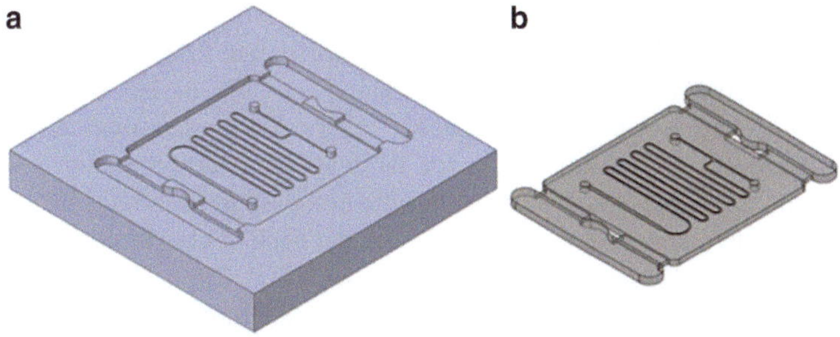

Fig. 4.7 Example mould tool (**a**) and injection moulded chip (**b**). Note that to fabricate inlet holes on the device an extruded pin is machined into the mould

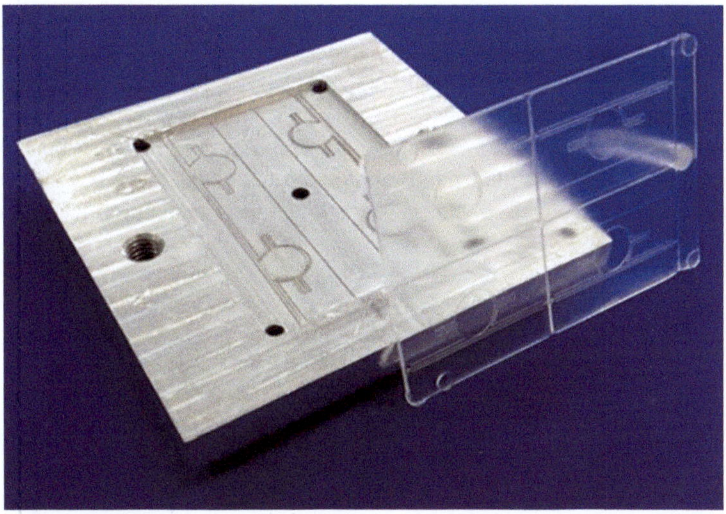

Fig. 4.8 Example aluminium injection mould tool and COC chip

Injection moulding has the additional benefit, when compared to alternative production methods, of raw material granules being able to manufacture devices of any size or shape. This limits material waste and storage requirements whilst increasing the in house capability for fabrication versatility (Becker and Locascio 2002; Chen et al. 2010; Mair et al. 2006).

Hot Embossing

Although not as wide ranging in terms of everyday application or consumer products, the process of hot embossing possesses similarities to injection moulding. As with injection moulding, this fabrication method first involves the production of an inverted mould including all the relevant features of the device. However, rather than melted material being forced into a mould, a work piece is stamped, thus imprinting the features onto the polymer. Heat and applied load are required for this method to be effective and must be determined for each mould and polymer to ensure features are accurately reproduced. Figure 4.9 diagrammatically demonstrates the process of hot embossing for fabrication of lab-on-a-chip devices. A stock polymer work piece is positioned in a hot embossing machine (Fig. 4.9a) then a heated machined embossing tool, containing the desired chip features, is pressed on to the material with a predetermined pressure (Fig. 4.9b). Heating causes the polymer to soften whilst the applied pressure leaves an imprint of the features on the work piece (Fig. 4.9c). Embossing moulds are typically fabricated from metals due to their thermal stability at high temperatures, but can also be fabricated in silicon or even polymers. Embossing manufacture rates can be increased through the use of rollers in place of a press. Polymeric material is continually passed under a roller die rather than being individually stamped, thereby reducing the overall time taken to manufacture the devices. Whilst this increases the production efficiently of devices, the start-up costs are typically much higher due to the specialist equipment required in producing the die.

It is not possible to directly compare hot embossing with injection moulding since both processes have their relative advantages and disadvantages; however, it is immediately obvious that hot embossing is inferior with regards to mass

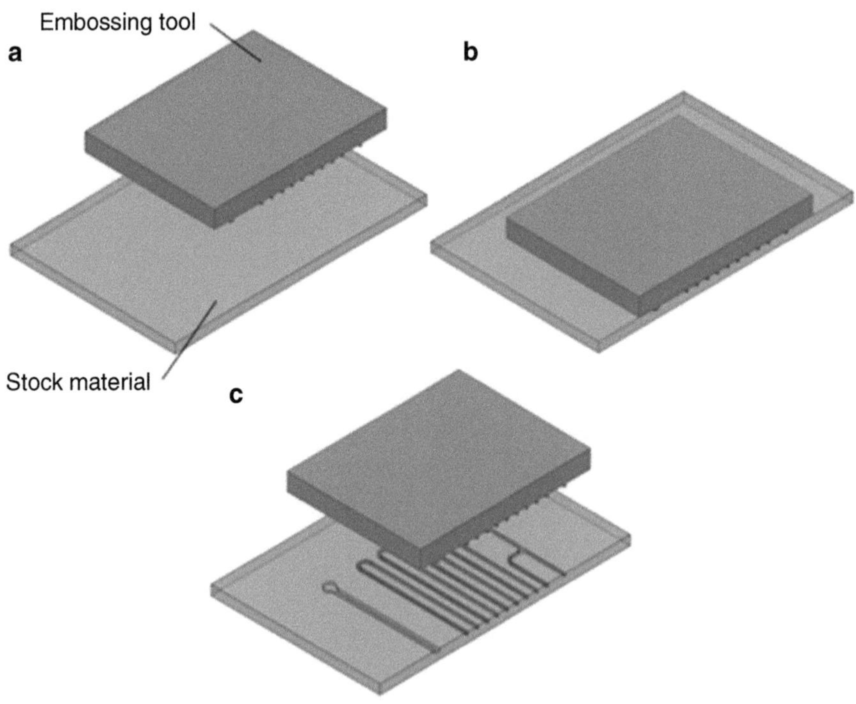

Fig. 4.9 Schematic example of the hot embossing fabrication process

production. Additionally, hot embossing does not present as much flexibility with regards to three-dimensional reproduction of features. This can lead to further processes being required to fulfil function requirement, for example holes cannot be directly moulded into the device. A single polymer work piece can be hot embossed by several different moulds, thereby enabling a range of shapes and features to be included on the device; however, this can be time consuming and create potential problems with alignment of features. Although careful consideration of stock work piece geometry can eliminate the majority of waste when utilizing hot embossing, it is necessary to purchase or manufacture the correct material shape and size for each device (Becker and Heim 2000; Chien 2006a, b).

3D Printing

Over recent years, 3D printing, also referred to as rapid prototyping, has become more prevalent in the production of components and consumer goods. Advances in this technology have transformed a previously specialist and expensive fabrication method into one that is potentially cost-effective and very rapid. The primary advantage of this technology is the ability to fabricate features and geometries that would otherwise be very difficult or require highly specialized equipment. Since the device is effectively 'printed' layer by layer, there are few limitations with regards to what the chip designer can manufacture. Minimum feature sizes are theoretically limited by the resolution of the machine and the 'layer thickness' that is applied, which some manufacturers claim can be as small as 16 μm (e.g. Objet Connex printer series).

Current 3D printing technology, with adequate resolution for the fabrication of microfluidic devices, is generally very expensive compared with alternative methods, primarily due to the costly raw material, and is therefore more commonly used for 'one off' production of experimental designs or radical new concepts. In such instances it is often cheaper and much more time effective to apply this process rather than manufacture a mould tool which may only be used to produce a few example devices. Continued refinement of the technology will likely see a reduction in material and machine costs thereby increasing the viability of this manufacturing method for lab-on-a-chip devices.

Whilst there appears to be great potential for the utilization of 3D printed devices it is important to recognize the limitations of this method. The range of polymers that can be fabricated is somewhat limited when compared to that of injection moulding or hot embossing. As a result of this reduction in possible fabrication materials, it is likely that factors such as mechanical integrity, thermal stability or chemical compatibility will be compromised when compared with alternative manufacture methods or materials. These compromises effectively limit the possible applications of devices fabricated using 3D printing (King et al. 2014; Kitson et al. 2012; Krejcova et al. 2014).

4.3.3 *Soft Polymers*

Soft polymers, also commonly referred to as elastic polymers have been widely used for the prototyping of microfluidic systems. This is primarily a result of a simple and relatively accessible manufacturing process. Unlike the high start-up costs associated with CNC micromilling of glass or injection moulding of rigid polymers, soft polymers offer a cost-effective method of experimenting with and producing devices with little expertise. Indeed, a large proportion (45–50 %) of the published works on microfluidic devices utilize this material and its associated fabrication method (Becker 2011).

It is possible to utilize a variety of polymers in this process; however, the vast majorities are fabricated from polydimethylsiloxane (PDMS). Mass production of soft polymer devices benefits from the use of mould tools which ensure repeatability of features and device geometries (see Fig. 4.10 for an example of this process). There are, however, limitations with regards to the rate at which a device may be produced due to the curing time of the material.

Relative advantages and disadvantages of soft polymers are listed as a summary below.
Advantages

- Excellent material for rapid and cost-effective manufacture of prototype devices.
- Integration of micro-valves, micro-pumps and other features into the devices is possible due to the flexibility of the material.
- Accessible and relatively low start-up costs.
- Moulds may be manufacture using a variety of materials since they are not subject to excessive heat or pressure.
- Three-dimensional and high aspect ratio features can be manufactured from this material.

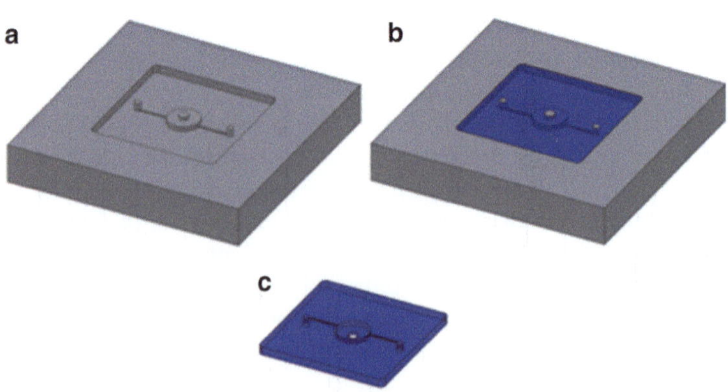

Fig. 4.10 Schematic example of cure moulding fabrication process

Disadvantages

- Possible fabrication methods are limited due to the elasticity of the material (e.g. it is not possible to micro-machine soft polymers).
- Fragility of devices due to their inherent softness.
- Curing time limits the capability of PDMS devices to be effectively mass produced.
- PDMS is hydrophobic. This is generally an undesirable property when attempting to manipulate fluids. Surface treatment may be required as an additional process to alter properties.
- Relatively limited chemical compatibility and surface adsorption of molecules.
- Swelling of the material when in contact with solvents can result in deformation of channels and feature geometry.

Cure Moulding

Cure moulding of PDMS is somewhat similar to the injection moulding of rigid polymers. Firstly, a mould of the desired channels and features is produced from a metal or other stock material (micromilled polymer) (Fig. 4.10a). The requirement for a metal mould is negated when cure moulding due to the relatively low temperatures and pressures involved in this process. As such, this fabrication process has the advantage of alternative manufacture methods being implemented in the production of a mould (e.g. creation of a glass mould through etching). Rather than softening a material with a high temperature as with injection moulding, PDMS is poured into a mould and allowed to cure (Fig. 4.10b), usually at an elevated temperature (around 60–70 °C), for up to 24 h. Once cured, the chip is removed from the mould (Fig. 4.10c) (Friend and Yeo 2010; McDonald and Whitesides 2002).

4.4 Assembly

Manufacture of channels and features into glass or polymer materials generally represents only part of the fabrication process. The majority of lab-on-a-chip devices are comprised of two (or more) pieces that are assembled to form a functional device. The design of optimized devices will consider the manufacturing implication of each piece and design them so as to maximize production efficiently. The simplest device is likely to involve the assembly of one plate containing the features and inlet/outlet holes with a sealing plate to prevent leakages and ensure the flow can be manipulated through the desired channels. Sealing the device is also critical for chemical and biological applications to ensure there is no contamination of the reagents. Depending on the requirements of the device, it may be necessary, or indeed preferable, to construct the chip from multiple feature-containing layers.

For instance, channels in one layer that match up to a cover containing the inlet/outlet holes. Although this can complicate the assembly process to some extent, due to the requirement for layer alignment, it has the potential to further enhance functionality of the chip.

Material selection will limit the number of applicable assembly methods (e.g. ultrasonic welding cannot be undertaken on a glass chip) and must therefore be considered as part of the design and fabrication process. In order to accommodate the physical functionality of the device throughout the assembly process the geometry of channels and features must not be distorted (e.g. blockage of channels). Compatibility of reagents with possible assembly methods may also need consideration, for example, adhesives may not be suitable when the flow of aggressive solvents is required.

4.4.1 Adhesives

Adhesively bonded chips may be assembled at room temperature and do not require specialist equipment, making it a desirable method for both prototyping and potentially cost-effective mass production though the use of automation. Adhesive bonding typically considers the use of glues in the assembly process. Liquid adhesive is generally not preferable due to difficulties in ensuring it does not enter, and thus block, the microfluidic channels. Additionally, liquids spread when a clamping pressure is applied, further increasing the likelihood of channel blockage. Adhesives with increased viscosity can be utilized through the use of tapes or other self-adhesive materials. Double-sided tapes also enable the adhesion of multiple chip layers. These materials still have the potential to block channels, especially when they are applied with a high pressure across channels with a low aspect ratio; however, relatively rigid backing material and a consideration of application methodology limit this issue.

Further fabrication stages may be required when assembling a chip which features multiple layers to ensure that fluids can be directed from one layer to another. This could simply entail manually cutting access holes in the adhesive to ensure a continuous fluid flow. Alternatively, a laser cutter may be used or an intricate die may be designed to transfer bespoke patterns onto adhesive layers.

Adhesive bonds vary between materials due to their relative surface energy. Materials with high surface energy, such as metals or glass form a stronger bond than those with lower surface energy such as polymers. Although the bond may be weaker when applied to polymers it is not necessarily unsuitable, however, it is important to consider the relative limitations of this assembly method. Weaker adhesion can lead to an increase in potential for device failure though leaking or delamination, thus it is important to consider the service requirement of the device (operating pressure and temperature, etc.) prior to the use of adhesives. Regardless of material, correct preparation of adhesion surfaces is required in order to maximum the bond strength. Typically, surface preparation involves degreasing, washing and drying the surfaces in addition to ensuring that they are flat and free from defects.

Adhesives have the additional benefit of being able to bond two or more different materials together. This aids in the possible development of hybrid devices that utilize the relative advantages of the various possible raw materials or fabrication methods (Kim et al. 2009; Martinez et al. 2008; Tsao and DeVoe 2009).

4.4.2 Glass Fusion

Glass plates may be bonded together though the application of a pressure whilst in a high temperature environment, usually a furnace. Following surface treatment of both glass wafers, they are placed in a furnace which is gradually heated to around 600–650 °C (below the glass transition temperature). Bonding is facilitated by a chemical reaction between hydroxide groups on the glass surfaces which when heated from covalent siloxane bonds between the two wafers (Daridon et al. 2001; Stjernstrom and Roeraade 1998).

4.4.3 Thermal Bonding

Compression of two polymer surfaces at elevated temperatures results in melting of the surface layers that are subsequently fused together upon cooling. Deformation of features though softening can occur in the device during the process, in particular those that are subject to pressure during heating. It is also important to ensure that excessive pressure, which could lead to channel blockage, is not applied during this process.

This assembly method can result in a very strong bond relative to the material strength, as such; it is typically able to remain functional under high pressures. Device compatibility with temperature or aggressive reagents is also less of a concern providing the initial material is suitable for the application. This assembly method also eliminates the need for an intermediate layer (as with adhesion) (Becker and Locascio 2002; Tsao and DeVoe 2009).

4.4.4 Ultrasonic Welding

Polymers can be fused together by melting microstructural surface features through the application of ultrasonic vibrations. These features, referred to as energy directors, are usually triangular in cross-section in order to focus the vibrational energy on a small area, causing them to melt. Simultaneously applying a compressive force causes the melted polymer to bond to its opposite layer upon cooling.

Location and geometry of energy directors is critical to the successful bonding of polymer layers since a sufficient amount of material must be melted to create a strong bond. However, if a feature is too large, there is a risk of the melted polymer entering nearby channels or holes and causing blockages. Unlike adhesive and thermal bonding, ultrasonic welding is localized on the chip, thus, it may only be possible to weld around channels and features and not across the entire device. As a result, the bond between layers may not be as strong as alternative methods, increasing the importance of considered energy director design.

In order to efficiently utilize ultrasonic welding as an assembly method, energy directors must be incorporated into the design process, for example, when designing and manufacturing a mould tool for injection moulding (Truckenmuller et al. 2006; Tsao and DeVoe 2009).

4.4.5 Contact Bonding

Contact bonding of soft polymers is a commonly applied and effective method of bonding PDMS devices. This assembly method has the added advantage of being able to assemble the PDMS chips with other materials such as glass, thereby enabling hybrid lab-on-a-chip device to be fabricated.

PDMS is strongly hydrophobic and as such will not adhere to other materials. It is therefore necessary to treat the PDMS with oxygen plasma in order to modify the exposed surface. Within one minute of this treatment, the device is placed against a cover layer, forming a permanent bond. It is possible to delay this adhesion, for example to aid in the alignment of devices, for a short time by using a small quantity of methanol (Eddings et al. 2008; McDonald et al. 2000).

4.5 Discussing Fabrication Possibilities

The previous sections have discussed the relative advantages and disadvantages of fabricating lab-on-a-chip devices using various manufacture and assembly methods. There has been an emphasis on the relative advantages of each material and process as well as considerations regarding the achievable fabrication tolerances. Achievable tolerances of a device are not necessarily the same as possible manufacturing or assembly accuracy. For example, a CNC machine with a positional accuracy of 10 μm may machine a glass wafer to within 100 or 200 μm precision due to limited tool tolerance (due to coarse diamond burs) and potential work piece damage during the process. However, the same machine may achieve close to the 10 μm precision when milling a metal mould tool due to the higher resolution of the work piece relative to glass. This section will compare and discuss the relative advantages and disadvantages of the most common lab-on-a-chip manufacture methods and the relative achievable tolerances associated with them.

Additionally, assembly methods are considered and the achievable accuracy for device alignment is considered.

4.5.1 Prototyping

Production of prototype devices is essential for determining the viability of new chip configurations as well as further optimizing the functionality or performance of existing devices. This is universal, regardless of whether the chip will be used solely in a research environment or considered for mass production and commercialization. As well as providing reliable data regarding the output from on-chip processes, a prototype device would benefit from being fabricated as quickly as possible. This is particularly important when a number of designs require testing or iterative design changes are necessary to refine a device.

Direct micromilling of stock material (e.g. glass or polymer) is relatively advantageous for prototyping since this method minimizes the time taken to produce the device. Additionally, it may be possible to modify it in some way after the initial fabrication (widening channels, incorporating additional access holes, etc.). Processes that require the manufacture of moulds (e.g. injection moulding or embossing) typically require a greater time commitment and more resources since they involve an additional production stage. Production of effective chips is also somewhat dependent on the optimization of moulding/embossing parameters that can also be time consuming. Despite the necessity for a moulding process, soft polymers are commonly utilized for prototyping, especially in academic research. Whilst this may not be considered an ideal method when assessing alternatives, the low cost and relative simplicity of cure moulding make it appealing for environments that do not have access to precision machines or the necessary expertise to implement them effectively.

Producing large quantities of chips is generally not required at the prototyping stage thereby negating the primary advantage of processes such as injection moulding; however, it is important to consider the long term requirements for chip fabrication. Should large quantities ultimately be required (e.g. commercialization of product), the necessary processes and functions taking place on prototype devices must also be compatible with chips fabricated through a method suited to mass production. For example, processes that are successful in a glass prototype device may not be transferable to one made from injection moulded PMMA due to relative incompatibility of aggressive reagents or thermal requirements. This further emphasizes the requirement for fabrication methods to be considered during the design stage.

Whilst the requirement for high precision in prototype devices may not appear to be a primary concern, it is critical for the accurate assessment of chip success that the features are representative of the initial design. Required tolerances will be highly dependent on chip function or application and could vary from simply satisfying the flow criteria to validating an intricate arrangement of channels and features across multiple plates. It is advantageous to consider the required device tolerances during the chip design stage thereby enabling the manufacture selection process to take this into account along with the most suitable prototyping methods.

4.5.2 Mass Production

Assessing the potential for mass production of chips is not necessary for all devices or applications, indeed considering such factors could be deemed detrimental to the design and development of some research chips. It is inevitable that some compromises are made when attempting to mass produce a device, for example, the chemical compatibility of injection moulded polymer devices relative to glass. It may be that mass production of a device requires alternative on-chip operations to be considered in order to facilitate compatibility in a polymer device.

It is clear from considering applicable fabrication methods that the most effective mass production technique is injection moulding. This is evidenced by the continued successful application of injection moulding across the manufacturing industry. Application of this method to lab-on-a-chip devices results in vastly increased production rates and reduced unit costs, thereby creating potential for product commercialization. Adoption of this technology is based on a number of factors, one of which is the relative affordability of the devices. As such, injection moulding is key to the further integration of lab-on-a-chip devices into everyday applications.

Chips requiring operations that are incompatible with polymers (e.g. those utilizing solvents or high temperatures) may not be mass producible though traditional means. It has already been stated that the functionality and ability to perform the necessary operations remains key to the success of any lab-on-a-chip device. As such, it may be essential to compromise on the material or fabrication process in order to accommodate the functional requirements of the chip. In this case, appropriate fabrication processes would require optimization to maximize the production efficiency. For example, should glass be the only viable option for device fabrication, the use of large wafers, from which multiple devices can be produced, and simultaneous production processes would increase the manufacturing output and reduce the associated costs. It is likely that the large scale production of any device though unconventional means will have a higher unit cost than one produced using commonly adopted techniques. Addressing the production requirements and any compromises that may be necessary during the chip design stage is vital for ensuring a successful outcome for the device, especially when producing a commercial product.

Producing large quantities of devices alters the chip tolerances that are acceptable since there is likely to be inherent variations between batches. In the case of injection moulding, the manufacture of a high quality mould (ideally from a hard metal such as steel) using CNC machining should achieve a precision of ± 10 μm or less. Production of chips using this mould should exhibit a similar level of precision; however, this is reliant on the successful optimization of moulding parameters and their continuation between batches. Deviation from these parameters may result in distortion of the moulded features as a result of flaws such as sink marks or preferential shrinkage. Necessary quality control measures would need to be implemented to assess an appropriate sample of devices from each batch to ensure they are within required geometric tolerances. Damage or degradation of the mould tool can occur during production which will likely result in the manufacture of flawed devices, thereby increasing the importance of implementing adequate quality control measures. The importance of quality control is exacerbated when atypical methods, which have greater potential for material variation and manufacturing flaws, are used for large volume production.

4.5.3 Assembly

Regardless of the precision achieved during the chosen manufacture processes, assembly of devices has considerable potential for fabrication inaccuracies through misalignment. Assembly inaccuracies may be dependent on the chosen method, for example, the use of adhesives or contact bonding could potentially allow for sliding of one component relative to another. Whilst it is conceivable that certain methods are more susceptible to misalignment than others, the use of assembly jigs or considered alignment features somewhat limits the relative potential for inaccuracies in each process. The consequences of misalignment during device assembly vary depending on the scale of features and the functional requirements but could potentially range from impaired performance to failure of the device.

Whenever possible, the chip designer should look to eliminate, or at least reduce, the requirement for rigid tolerances in devices. For example, fabrication of all channels and access/outlet holes into a single wafer eliminates the requirement for precise alignment during the assembly process. For such a device, the achieved precision would be entirely dependent on the geometrical accuracy of the implemented manufacturing processes in addition to any related inaccuracies associated with integration of two or more methods. Whilst the application of certain materials of methods could lead to relatively significant inaccuracies, utilizing techniques that have high positional precision (e.g. wafer dicing and CNC micromilling) will minimize these and increase the likelihood of the device fulfilling the initial functional criteria.

For some chips it may not be possible to simplify the design or manufacture process in such a way so as to reduce the necessary geometrical precision, especially when fabricating high performance or increased functionality devices. In such scenarios, it is critical that the achievable tolerances for the utilized manufacture processes are considered and assessed. Utilizing multiple layers or integrating various components further increases the potential for misalignment and with it the importance of applying effective alignment techniques during device assembly. Alignment of a glass device during the assembly process could be achieved though considered placement and CNC micromilling of features, such as holes, into the wafer or by using diced edges. The process of determining the most suitable assembly reference for such a device would first consider the positional accurate of the relevant machines. Providing the machines are of relatively high quality and well maintained, it is conceivable that they will both work to a positional tolerance of around 10 μm. Although comparable in terms of quoted accuracies, the achievable precision for two such processes will likely differ due to factors relating to the work material and machining tools. Micromilling of glass requires the use of diamond coated end mill tools which, due to their relatively large diamond burs, may be more susceptible to machining inaccuracies than the coating on very fine dicing blades. Additionally, the process of micromilling features is more likely to induce flaws (e.g. chipping and cracking) that sectioning wafers. This could potentially amount to a feature inaccuracy that is an order of magnitude greater than that of the diced edge. Considering the potential for relative inaccuracies during the fabrication, it would be more suitable to use diced edges for assembly alignment rather than milled holes. The conclusion drawn from this example are based on assumptions and is somewhat generalized, therefore, it may not be applicable to every material or manufacturing environment. However, it does serve to highlight the importance of identifying the relative inaccuracies encountered during the manufacturing process and ensuring the most suitable feature (e.g. hole or edge) is utilized for alignment during device assembly.

Potentially, polymer devices have greater scope for ensuring alignment features can be integrated with greater accuracy. Utilizing injection moulding or embossing as a production method relies on the creation of a mould tool, typically though micromilling of metal work pieces. Milling of metals generally enables a high degree of accuracy due to the relative achievable resolution. As such, it is possible to precisely mould or emboss alignment features which can then be matched with corresponding features on the opposite plate. Inaccuracies associated with machined features are limited to the positional tolerance of the equipment (e.g. CNC machine). When polymers are heated or cooled there is a risk of shrinkage or warping of the material, which could lead to distortion of features. Whilst this could contribute to the misalignment of chips, steps can be taken to correct inaccuracies though careful mould design or consideration of the moulding parameters.

References

Abgrall P, Gue AM (2007) Lab-on-chip technologies: making a microfluidic network and coupling it into a complete microsystem—a review. Journal of Micromechanics and Microengineering 17(5):R15–R49

Becker H (2011) Famous last words. Lab Chip 11(13):2133–2134

Becker H, Heim U (2000) Hot embossing as a method for the fabrication of polymer high aspect ratio structures. Sens Actuator A-Phys 83(1–3):130–135

Becker H, Locascio LE (2002) Polymer microfluidic devices. Talanta 56(2):267–287

Chen Q, Li G, Jin QH, Zhao JL, Ren QS, Xu YS (2007) A rapid and low-cost procedure for fabrication of glass microfluidic devices. Journal of Microelectromechanical Systems 16 (5):1193–1200

Chen CS, Chen SC, Liao WH, Chien RD, Lin SH (2010) Micro injection molding of a microfluidic platform. Int Commun Heat Mass Transf 37(9):1290–1294

Chien RD (2006a) Hot embossing of microfluidic platform. Int Commun Heat Mass Transf 33 (5):645–653

Chien RD (2006b) Micromolding of biochip devices designed with microchannels. Sens Actuator A-Phys 128(2):238–247

Daridon A, Fascio V, Lichtenberg J, Wutrich R, Langen H, Verpoorte E, de Rooij NF (2001) Multi-layer microfluidic glass chips for microanalytical applications. Fresenius Journal of Analytical Chemistry 371(2):261–269

Eddings MA, Johnson MA, Gale BK (2008) Determining the optimal PDMS-PDMS bonding technique for microfluidic devices. Journal of Micromechanics and Microengineering 18(6)

Friend J, Yeo L (2010) Fabrication of microfluidic devices using polydimethylsiloxane. Biomicrofluidics 4(2)

Ghobeity A, Papini M, Spelt JK (2009) Abrasive jet micro-machining of planar areas and transitional slopes in glass using target oscillation. Journal of Materials Processing Technology 209(11):5123–5132

Iliescu C, Chen BT, Miao J (2008) On the wet etching of Pyrex glass. Sens Actuator A-Phys 143 (1):154–161

Iliescu C, Taylor H, Avram M, Miao JM, Franssila S (2012) A practical guide for the fabrication of microfluidic devices using glass and silicon. Biomicrofluidics 6(1)

Khan AA, Haque MM (2007) Performance of different abrasive materials during abrasive water jet machining of glass. Journal of Materials Processing Technology 191(1–3):404–407

Kim J, Surapaneni R, Gale BK (2009) Rapid prototyping of microfluidic systems using a PDMS/polymer tape composite. Lab Chip 9(9):1290–1293

King PH, Jones G, Morgan H, de Planque MRR, Zauner KP (2014) Interdroplet bilayer arrays in millifluidic droplet traps from 3D-printed moulds. Lab Chip 14(4):722–729

Kitson PJ, Rosnes MH, Sans V, Dragone V, Cronin L (2012) Configurable 3D-Printed millifluidic and microfluidic 'lab on a chip' reactionware devices. Lab Chip 12(18):3267–3271

Krejcova L, Nejdl L, Rodrigo MAM, Zurek M, Matousek M, Hynek D, Zitka O, Kopel P, Adam V, Kizek R (2014) 3D printed chip for electrochemical detection of influenza virus labeled with CdS quantum dots. Biosensors & Bioelectronics 54:421–427

Mair DA, Geiger E, Pisano AP, Frechet JMJ, Svec F (2006) Injection molded microfluidic chips featuring integrated interconnects. Lab Chip 6(10):1346–1354

Martinez AW, Phillips ST, Whitesides GM (2008) Three-dimensional microfluidic devices fabricated in layered paper and tape. Proceedings of the National Academy of Sciences of the United States of America 105(50):19606–19611

McDonald JC, Whitesides GM (2002) Poly(dimethylsiloxane) as a material for fabricating microfluidic devices. Accounts of Chemical Research 35(7):491–499

McDonald JC, Duffy DC, Anderson JR, Chiu DT, Wu HK, Schueller OJA, Whitesides GM (2000) Fabrication of microfluidic systems in poly(dimethylsiloxane). Electrophoresis 21(1):27–40

Nouraei H, Kowsari K, Spelt JK, Papini M (2014) Surface evolution models for abrasive slurry jet micro-machining of channels and holes in glass. Wear 309(1–2):65–73

Park DS, Cho MW, Lee H, Cho WS (2004) Micro-grooving of glass using micro-abrasive jet machining. Journal of Materials Processing Technology 146(2):234–240

Sein H, Ahmed W, Jackson M, Woodwards R, Polini R (2004) Performance and characterisation of CVD diamond coated, sintered diamond and WC-Co cutting tools for dental and micromachining applications. Thin Solid Films 447:455–461

Stjernstrom M, Roeraade J (1998) Method for fabrication of microfluidic systems in glass. Journal of Micromechanics and Microengineering 8(1):33–38

Truckenmuller R, Ahrens R, Cheng Y, Fischer G, Saile V (2006) An ultrasonic welding based process for building up a new class of inert fluidic microsensors and -actuators from polymers. Sens Actuator A-Phys 132(1):385–392

Tsao CW, DeVoe DL (2009) Bonding of thermoplastic polymer microfluidics. Microfluidics and Nanofluidics 6(1):1–16

Whitesides GM (2006) The origins and the future of microfluidics. Nature 442(7101):368–373

Yan BH, Wang AC, Huang CY, Huang FY (2002) Study of precision micro-holes in borosilicate glass using micro EDM combined with micro ultrasonic vibration machining. Int J Mach Tools Manuf 42(10):1105–1112

Yetisen AK, Akram MS, Lowe CR (2013) Paper-based microfluidic point-of-care diagnostic devices. Lab Chip 13(12):2210–2251

Zhou M, Ngoi BKA, Yusoff MN, Wang XJ (2006) Tool wear and surface finish in diamond cutting of optical glass. J Mater Process Technol 174(1–3):29–33

Zong WJ, Li ZQ, Sun T, Cheng K, Li D, Dong S (2010) The basic issues in design and fabrication of diamond-cutting tools for ultra-precision and nanometric machining. Int J Mach Tools Manuf 50(4):411–419

Chapter 5
Fluidic Platforms and Components of Lab-on-a-Chip devices

Christiane Neumann and Bastian E. Rapp

Abstract In recent years the distribution of Lab-on-a-chip devices as well as micro-total analysis systems in applications such as analytical chemistry, biochemistry, biotechnology, microsystems technology, or clinical diagnostics has increased significantly. In order to allow multiple assays to be carried out on this devices components that enable fast tests and quantitative measurements are needed. The first systems which fulfilled these requirements were paper based devices. The development of these systems was based on chromatographic techniques. The basic principle is already known as so termed spot tests since the 1930s. The trend to take more and more applications out of the laboratory to the user started the development of a large number of platforms for point of care devices. These platforms can be driven by different ways, e.g., pressure, capillary flow, or electro kinetic effects. Complex applications need additional fluidic components such as pumps, valves, sensors, or mixers. In this chapter different fluidic platforms as well as fluidic components will be described. Applications of platforms and integrated components are exemplarily demonstrated by means of case studies.

5.1 Introduction

Since the dawn of analytical chemistry, designated tests with simple (semi)quantitative readouts and short sample-to-result intervals have been of interest both scientifically and commercially. Early work on the so-called Tüpfelreaktionen (German for spot analysis) has been conducted as early as 1935 and several notable contributions to this field have been made, among others by Feigl (Feigl 1935) and Winkley and coworkers (Winkley et al. 1937) decades ago. Originally motivated by the work on separation-based chromatography systems on paper, the need for simple and robust tests which could be carried out outside of the laboratory, close to, e.g., the bedside of a patient has motivated numerous developments over

C. Neumann • B.E. Rapp (✉)
Institute of Microstructure Technology (IMT), Karlsruhe Institute of Technology (KIT), Hermann-von-Helmholtz-Platz 1, 76344 Eggenstein-Leopoldshafen, Germany
e-mail: bastian.rapp@kit.edu

J. Castillo-León, W.E. Svendsen (eds.), *Lab-on-a-Chip Devices and Micro-Total Analysis Systems*, DOI 10.1007/978-3-319-08687-3_5

decades on the so-called point of care (PoC) devices. With the emergence of microfluidics as a scientific discipline, the PoC concept has gained new momentum and has been an active field of development in various scientific disciplines ranging from analytical chemistry, microsystem engineering, chemical biology, biotechnology, clinical analytical chemistry as well as clinical diagnostics.

The first analytical systems based on microfluidics known from literature date back to the seminal work of Terry et al. in 1979 which described an integrated gas chromatography system and analyzer which was manufactured on two silicon wafers (Terry et al. 1979). The system allowed separation and detection of a gaseous mixture of organic solvents and featured an anemometric detector, a passive adsorber column and integrated monolithic microvalves. This system was a brief outlook on what integrated analytical systems could achieve. Later the concept was formulated, first in 1983 by Widmer in his seminal paper in which he predicted that "such sophisticated, integrated systems, characterized by their exchangeable modular setup, will have widespread future applications in industry. They will be part of the analytical impact which is changing the face of the chemical and allied industries" (Widmer 1983). In this work, Widmer already defined the term "miniaturized total analysis systems (μTAS)." The term was precisely described and discussed by Manz et al. in 1990 which marks the beginning of large-scale research activities and the start of Lab-on-a-Chip (LoC) device development. The fact that LoC devices have become important players on the healthcare market and analytical chemistry can be studied in Chap. 1 on the historical development of LoC and commercial success stories found in the work at hand or in the literature (Chin et al. 2012) to which we would like to refer the reader.

As the name suggests, LoC systems are designed to emulate a full-scale analytical laboratory with a very narrow analytical focus. In sacrificing universality, a significant gain in compactness and simplicity of operation can be expected. This is a significant advantage of LoC systems compared to simply making classical analytical instruments (such a, e.g., a gas chromatography or a mass spectroscopy instrument) portable. The readout of a LoC system is simpler, the intervals required for obtaining the readout (often referred to as sample-to-result or sample-to-answer interval) is significantly shorter and the amount of media (sample, consumables, etc.) required for the test is usually smaller because of the downsizing effect. For many test scenarios (such as bedside monitoring or point-of-care testing) these effects usually significantly outweigh all disadvantages that LoC systems may be (rightfully or not) associated with. One of the most prominent arguments always brought forth against LoC technology is the fact that tailored assays and devices need to be developed for each assay required. In essence, this is true. However, one of the main advantages of LoC devices is the fact that they consist of a discrete number of functional building blocks which are arranged such that they fulfill the task at hand. It is therefore essentially a question of finding, optimizing and properly arranging these building blocks which are the essence of LoC system design.

It is the aim of this chapter to describe a selection of the most important functional building blocks and highlight how and why they are used in LoC systems. Depending on their function these components may be created monolithically from a single material (such as a polymer) or may consist of a multitude of materials as is usually the case for sensors. Naturally, the type of material in question defines the choice of suitable manufacturing technology. The development of LoC systems has and always will be closely linked to manufacturing technologies. Early microfluidic devices have originated from structuring and processing techniques developed for classical microelectromechanical systems (MEMS) and the first devices published were therefore mainly created from silicon and glass. Early reviews on manufacturing technology such as the excellent paper by Gravesen et al. written in 1993 (Gravesen et al. 1993) summarize these processing methods. More recent discussions of manufacturing techniques can be found, e.g., in Waldbaur et al. (Waldbaur et al. 2011) and in Chap. 4 found in the work at hand.

In the following, we will discuss the most important components of LoC systems. We will start by discussing the choice of sampling liquid and the various means of introducing the samples into the systems. Next we will briefly highlight the most important microfluidic platforms commonly used in LoC systems as well as the most important building blocks these systems are commonly created from. Naturally the choice of platform will strongly influence the choice of components. The discussion will conclude with a selected number of examples from the literature that allow studying the conceptual setup of LoC systems from these building blocks.

5.2 Lab-On-a-Chip Platforms and Components

5.2.1 Sample Introduction Onto Chip

There are a lot of potential sample fluids which can be analyzed with LoC systems. A great percentage of applications deals with the so-called biofluids, which are samples obtained from the human body. Typical examples include blood plasma, whole blood, urine, tissue fluids, or cerebral spinal fluid (Crevillén et al. 2007). Besides biofluids LoC systems may also be developed for environmental analysis, e.g., using tap water, mineral water, river water, groundwater, smoke particles or electroplating sludge, as well as for food analysis, using e.g., milk, red wine, juices, green tea extract, vegetables or fruits (Gravesen et al. 1993).

Especially for the analysis of biofluids a relative large amount of the sample (within the range of hundred microliters up to few milliliters) is required in order to obtain a few microliters for analysis post after sample preparation (Lloyd 1996). Various means of sample preparation have been described in the literature. One commonly encountered step is separation of the analyte from a biological matrix.

This can be done via liquid-liquid or solid-phase extraction (Lloyd 1996; Lichtenberg et al. 2002), physical particle filtering and retention by integrated flow restrictions or by diffusion in laminar flows (Lichtenberg et al. 2002). Another option is usage of semipermeable membranes in order to separate the sample stream from a perfusion liquid resulting in a microdialysis setup (Lichtenberg et al. 2002). Another commonly found alternative is separation using free-flow electrophoresis, a technique that uses electrokinetic movement in an electric field arranged perpendicular to the flow direction (Lichtenberg et al. 2002). Cells are usually lysed in order to obtain analytically relevant samples. Potential means of doing so include chemical lysis, thermal lysis, lysis under ultrasound as well as electroporation with subsequent sample purification (Lichtenberg et al. 2002). For some scenarios certain means of sample preconcentration may need to be implement, e.g., by using field amplification stacking, a technique that uses high electrical fields applied to an injected sample plug which results in anion and cation separation and preconcentration at the edges of the plug. Field amplified injection techniques use a similar mechanism predominantly at the sample-buffer interface during injection of the sample into LoC system (Lichtenberg et al. 2002). Further preparation steps may include chemical or biochemical derivatization of the sample, e.g., by fluorescence labeling or complexation. Pretreatments regularly performed especially in bioanalytics include amplification, e.g., of DNA by polymerase chain reaction (PCR) or enzymatic digestion of DNA or proteins (Lichtenberg et al. 2002) prior to analysis in mass spectroscopy or separation in gel. In general it is advantageous to remove particles from the sample via extraction centrifugation and filtration (Crevillén et al. 2007). Some of these preparation methods like filtration and extraction (see below) can also be done on-chip.

After sample preparation the analyte has to be transferred into the LoC system. For simple capillary flow tests the sample can be directly dipped onto the sample pad and then passively transported via capillarity to a conjugation pad and the detection area (Haeberle and Zengerle 2007). The sample way also be provided into the system by suitable accessible wells or reservoirs located, e.g., in the cover plate of the system or via tubes or pipette tips glued onto the chip. These reservoirs can be manually loaded, e.g., with a syringe or pipette or with the help of a liquid handling station (Kopf-Sill 2002; Fredrickson and Fan 2004; Oh et al. 2005). If these reservoirs are designed accordingly, direct connection with a tubing system can be achieved. Sealing is usually accomplished via press fitting, flanging and bonding of these tubes to the LoC system. Several noteworthy commercial products exist that are designed accordingly to these principles. A commonly found example are Upchurch's Nanoports or LuerLok fittings which have been described, e.g., by Frederickson and Fan (Fredrickson and Fan 2004). These are simple methods to transfer the sample into the LoC system and they allow quick adaption to design changes which is why they are often chosen for prototyping. Disadvantages to consider include clogging of the channels due to penetration of adhesives into the channel network during assembly, high dead volumes and the time required for each connector individually which may sum up to hours of manual work if dozens of tubes need to be connected. Especially if the systems are designed for single use only, as is commonly the case in PoC testing, these concepts are only of limited use due to the manufacturing overheads. In addition the dead volumes of the individual connectors quickly sum up to inacceptable volumes of sample lost during the assay. In addition, sample contaminated due to contact with the adhesive used during assembly may be critical for sensitive applications especially when biological samples are used.

Another commercially available system for sample transfer into LoC systems is the so-called Sipper Chip from Caliper Life Sciences. These chips feature capillaries directly integrated into the planar LoC system structure for sample loading directly from a microwell plate via vacuum application (Kopf-Sill 2002; Chow 2002). Again, the capillaries have to be disposed if the chips are used as single-use component. Systems that allow reusing the fluidic interface typically rely on tubes or capillaries merely pressed against the LoC device temporarily via clamping or screwing (Fredrickson and Fan 2004). After sampling, the LoC device is disposed after releasing the temporary connection with the tubing system. The latter can be reused, potentially after cleaning. In order to achieve sufficient sealing, most of these devices use compressible sealing elements made from polydimethylsiloxane (PDMS) which are in direct contact with the sample (Saarela et al. 2006; Scott et al. 2013; Sabourin et al. 2009). In consequence, the chemical resistance and swelling behavior of the sealing element limits the range of potential liquids and samples which can be probed in the LoC system. Usage of flanged tubes avoids contact between sample and sealing element and allows, if used with a chemically inert tubing material such as polytetrafluorethylene (PTFE), probing of nearly every fluid resulting in a chemically inert, reversible, multichannel chip-to-world interface (Wilhelm et al. 2013a). Another means of introducing sample liquids into the LoC system is by using a dispensing platform which spots the liquid by applying short pressure pulses in a free-flowing manner from the dispenser directly into the LoC system. Here system contamination can be avoided entirely, albeit at the cost of more expensive instrumentation (Haeberle and Zengerle 2007).

An overview of the general principles for sample transfer into LoC system is shown in Fig. 5.1. For more information on the subject, the reader is referred to the article by Frederickson and Fan (Fredrickson and Fan 2004) which gives an overview about the different methods of world-to-chip interfacing.

Depending on the assay to be performed, a sample, once transferred into a LoC system, may need to undergo certain preparative steps before being analyzed. As such, the LoC system needs to provide suitable structures to hold and guide the liquid and to allow its proper manipulation. Certain components are usually employed for doing so. The choice of components is strongly influenced by the fluidic platform chosen for the LoC system. Also the choice of platform defines the layout of the assay to be performed. Several notable fluidic platforms have been described and established over the years. The following section will highlight the most important ones.

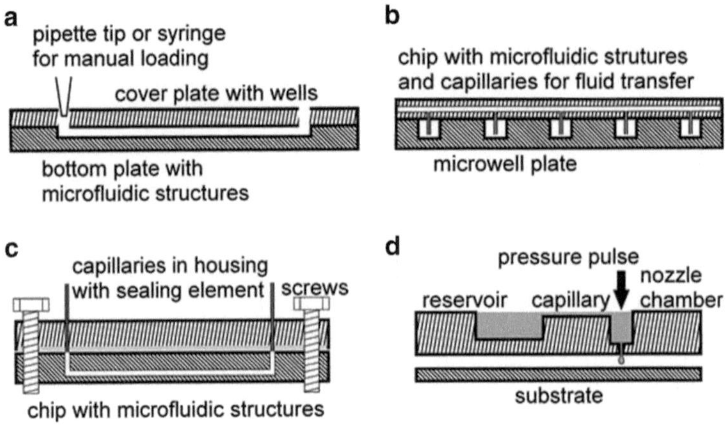

Fig. 5.1 General principles for transferring sample fluid to LoC systems. (**a**) Manual loading of wells or reservoirs in the cover plate of a LoC device. (**b**) LoC system with integrated capillaries for fluid transfer directly out of a microwell plate by vacuum application ("Sipper Chip," Caliper Life Sciences). (**c**) Connecting a LoC system with capillaries or tubes using a sealing element via clamping or screwing. (**d**) Dispensing system to transfer sample to a substrate or a LoC system via application of short pressure pulses. The nozzle chamber refills by capillary force from a reservoir

5.2.2 Assay Design and Fluidic Platforms

For the most part, LoC systems consist of components that provide physical means of guiding liquid, most commonly in form of microfluidic channel networks. The layout of these channel structures must be fixed during development of the LoC system and is usually not required to be changeable. However, depending on the complexity of the assay to be performed, certain means of manipulation may need to be incorporated, e.g., in form of microvalves that allow temporary blocking or unblocking of channel structures. Figure 5.2 shows a general scheme of a LoC device with different active and passive structures.

In addition to this rather classical approach several microfluidic platforms have been suggested over the years. The following section will briefly highlight the most important ones and discuss their significance for LoC devices.

5.2.2.1 Pressure-Driven Flow Platforms

Driving fluid flow by means of mechanical pressure is the most straightforward way of setting up a LoC device. These systems usually use valves for fluid control and pumps for creating fluid flow. As such, valves and pumps are the atomic building blocks of this type of platforms and numerous design concepts for building them have been described over the years. The most important actuation concepts will be summarized below. Depending on the actuator concept, pressure-driven platforms may lend themselves well to the creation of LoC systems. However, the need for

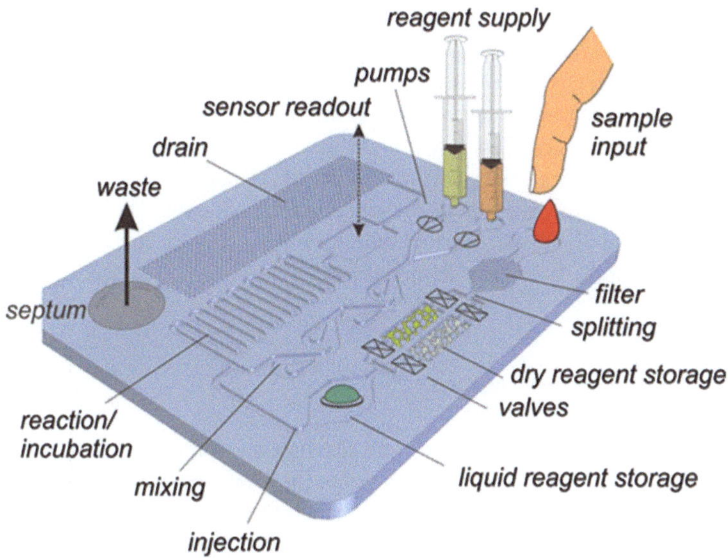

Fig. 5.2 Lab-on-a-Chip devices for point of care applications usually comprise a number of passive or active structures which are arranged such that a distinct application can be carried out with the chip at hand. Depending on the application such chips will contain means of introducing a sample liquid and/or reagents, potentially reagent storage on-chip, means of mixing, separating, or filtering of the fluids and structures for analytical readouts. Typically structures for storage of used reagents and waste liquids also need to be provided

creating driving pressure requires actuation which implies usage of a power source. This may limit the application range of these devices, depending on the complexity of the assay and the robustness of the analytical device. Pressure-driven flow platforms are the most versatile fluidic platform, especially if ample means of manipulation fluids are integrated. As such, these platforms are well suited for applications, in which the reaction to be carried is not yet fully characterized and the assay thus not completely fixed with all parameters yet. Precise fluid control allows exploiting inherent advantages of microfluidics such as laminar flow for separation of objects by deterministic displacement (Huang et al. 2004) or focusing of particles and thus sequential alignment in a microfluidic channel, e.g., for sorting of cells (Dongeun et al. 2005). Laminar flow has a number of interesting advantages such as controlled diffusion which allows for several interesting detection principles such as the microfluidic H-filter (Brody and Yager 1997) or T-sensor (Weigl et al. 1999).

Albeit their advantages, pressure-driven LoC systems are often deemed too difficult to set up and too complicated to operate. For on-site testing or PoC diagnostics, these systems usually rely on active (and thus expensive) components such as valves and pumps to be directly integrated into platform which makes the systems inherently complicated in setup and thus expensive on an industrial scale. Numerous pump and valve concepts have been described for this platform which will be summarized below. During the assay, these active components are contaminated with the analyte and thus must be disposed after each experiment. In order to reduce costs, numerous concepts have been suggested over the years which allow separating active (reusable) components from passive (disposable) components of the systems. These systems usually rely on indirect fluid handling as enabled, e.g., by indirect microfluidics using inert hydraulic oils (Rapp et al. 2009) or air (Sauer-Budge et al. 2009). However, these systems tend to become bulky in setup which is why usage is usually restricted to laboratory type assays for which numerous application examples have been demonstrated.

5.2.2.2 Capillary Flow Platforms

LoC assays based on capillary flow were the first microfluidic platform described and commercialized. In the form of test strips they have been and still are the gold standard for cheap diagnostics such as pregnancy tests. As the name suggests, these platforms rely on capillary flow induced usually by using structures with high internal volumes and small channels which absorb liquids conveniently. Typical examples include hydrophilic fleece structures or suitably structured porous monoliths. Often a small hydrophilic channel is sufficient for example sample delivery. Capillary flow platforms do not require pumps for producing flow which makes them especially suitable for low-cost applications or low resource settings. The most commonly employed readout concept in capillary flow platforms is optical readout. In binary tests, such as pregnancy, the readout is often a mere change of color visible to the human eye. If quantitative readout is required, a wide number of detectors and sensors can be readily incorporated into these platforms as is the cases, e.g., in the widely established blood glucose test strips. These assays rely on the electrochemical detection of glucose oxidation by surface immobilized enzymes such as glucose oxidase. The enzyme is immobilized on an electrode and the electrons freed during the oxidation process are collected by these electrodes creating a measurable current proportional to the blood glucose level. It has to be noted that passive capillary flow platforms are by far the most successful LoC platform. The market share of glucose sensing alone ranges in the billions of dollars annually and, with an estimated number of 300 million diabetes patients in 2025, will increase further (Newman and Turner 2005). No other PoC platform has had a comparable success story.

Interestingly, capillary flow platforms have recently regained significant scientific interest in the form of the so-called microfluidic paper-based analytical devices (μPADs). This developement was jolted by the introduction of paper-based microfluidics by the Whitesides group (Martinez et al. 2008). The principle concept relies on structuring a hydrophilic paper with sufficient capillarity such that parts of the paper are covered by a hydrophobic polymer (Martinez et al. 2007). More details on the manufacturing of μPADs can be found in the literature (Li et al. 2010) and in Chap. 7 found in this work.

Without a doubt, one can hardly image a platform more robust and simple in terms of operation than capillary flow platforms. They can be operated with virtually no instruments and if readout is provided by means of color changes, semiquantitative PoC devices can be set up without any power-consuming components. Therefore, capillary flow systems lend themselves well for the design of analytical instruments that fulfill the ASSURED criteria brought forward by the World Health Organization (WHO) Sexually Transmitted Diseases Diagnostics Initiative. ASSURED is an acronym of the criteria required of an instrument for robust operation in low-resource settings: affordable, sensitive, specific, user-friendly, rapid and robust, equipment-free and deliverable to end-users. Most of these requirements are well met by some of the assays described in the literature.

However, the capillary flow platform suffers from numerous disadvantages such as the difficulty in flow control as discussed. Another result of this is the fact that mixing is difficult in the mostly laminar flow regimes inside of the porous structures. Despite these problems capillary flow platforms will be an important platform to consider for future LoC applications.

5.2.2.3 Segmented Flow Platforms

The concept of segmented flow is taken from emulsion formation where small droplets of one phase (usually an aqueous phase) are dispersed into a second phase (usually a hydrocarbon or fluorinated oil). Segmented flow platforms make use of the fact that the oil/water interfaces (due to the difference in surface tensions) are instable and tend to break up if sufficient shear force is applied. Due to the nature of the emulsion, these systems are also commonly referred to as droplet microfluidic devices.

Oil/water emulsions are classically prepared, e.g., using porous membranes (Nakashima and Shimizu 1993; Suzuki et al. 1998) or by shear rupturing following extensive stirring (Mason and Bibette 1997). Another technique includes the so-called Rayleigh jet breakup (Berglund and Liu 1973) which exploits the Plateau–Rayleigh instability, a phenomenon which can be observed on falling water droplets even at the macroscopic length scale (Papageorgiou 1995). In 1997 a silicon-based system was proposed that allowed creation of emulsions using a micromachined microfluidic channel (Kawakatsu et al. 1997). This paper was the first to report creation of water-in-oil emulsions using microfluidic technology. This was later adapted to microfluidic devices using cross-flow for emulsification (Kawakatsu et al. 2000). Similar results can also be obtained using co-flowing streams in microfluidic systems (Umbanhowar et al. 1999). In 2001 a co-flowing microfluidic emulsification platform was demonstrated which used (for the first time) a polymer (in this case a light-curable polyurethane acrylate) as bulk material instead of capillaries or silicon (Thorsen et al. 2001). Today, the most commonly used materials for the creation of segmented flow platforms are glass (Funfak et al. 2007) and PDMS (Clausell-Tormos et al. 2008).

This platform has found wide adoption in the LoC community and has been used, among others, for PCR of DNA (Curcio and Roeraade 2002) and numerous cell based assays where the individual droplets have been used as small cell containers (Li et al. 2005). Due to the fact that droplet microfluidic systems usually require precision pumps for providing stable liquid flows in order to obtain droplets with sufficiently small size variations, droplet microfluidics has not found notable usage for out-of-the-lab applications. However, it is a noteworthy platform for numerous applications in clinical diagnostics and cell biology as well as for analytical chemistry (Teh et al. 2008).

5.2.2.4 Electrokinetics

Electrokinetic platforms range among the oldest LoC platforms described. In general, electrokinetics makes use of the fact that the individual dissolved components of a sample will move in an electric field according to their charge-to-mass ratio which defines the respective electrokinetic mobility. This type of mobility-dependent electrokinetic separation is usually referred to as electrophoresis and is one of the oldest analytical techniques used, especially in proteomics. It dates back to the work of German physicist Friedrich Kohlrausch who already noted that dissolved components in liquid can be separated by the application of electric fields (Kohlrausch 1897). Macroscopically, electrophoretic separation is usually applied inside of polyacrylamide gels, which is why the technique is referred to as "polyacrylamide gel electrophoresis" (PAGE). Gel electrophoresis has been used as a method in analytics since the late 1950s (Smithies 1955). As a PoC electrokinetic separation is usually carried out merely in buffer and referred to as capillary zone electrophoresis (CZE). Originally CZE was mainly performed in glass capillaries (Mikkers et al. 1979; Jorgenson and Lukacs 1981). In comparison with gels, the field strengths which can be applied when using capillaries are significantly higher. Electrophoresis has been among the first techniques transferred to a microfluidic platform and early commercial success stories, most notable the introduction of the one-dimensional gel electrophoresis system by Caliper (today Caliper Life Sciences) which is, in the form of the Agilent Bioanalyzer, still a successful product in the market (Goetz et al. 2004).

Besides the movement of charged particles by externally applied electric fields, small capillaries (such as microfluidic channels) also exert a secondary effect on the bulk liquid which results in lateral movement. This effect is referred to as electroosmotic flow (EOF). The general principle of EOF is shown in Fig. 5.3. The cause of this effect is the fact that charging a solid surface will result in the formation of an inversely charged second layer in the surrounding liquid and thus to the formation of a double layer. This double layer acts as a sliding bearing and allows (once an electric field is applied along the axis of the capillary or the channel) lateral movement with very flat and homogenous flow profiles across the channel. Bulk movement is facilitated by fluid viscosity coupling which ensures that the

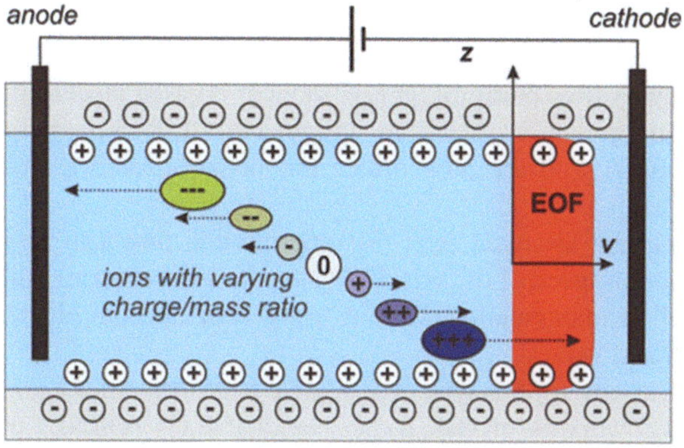

Fig. 5.3 Concept of electrokinetic movement inside of microfluidic channel. The example shows a negatively charged channel, e.g., when using a fused silica microchannel. Applying an electric field along this channel will result in ion movement inside of the microfluidic channel. This can be used for example for separation and analysis, e.g., by electrophoresis. However, this effect can also be used for bulk pumping of liquids to the electroosmotic flow

once the fluid layers in direct contact with the double layer are displaced, the bulk liquid is dragged along.

EOF and electrophoretic separation in conjunction can be used conveniently to separate almost all charged and non-charged components of a sample. In addition, EOF can also be used to move liquids inside of fluidic channels. As such, electrokinetic platforms can be considered one of the most versatile LoC platforms. They are operationally simple to set up as they only require immersing electrodes into the liquids and applying electric fields. Separation is convenient to implement and even detection (as in the case of for example liquid conductivity sensing) can be implemented with little to no instrumental overhead.

5.2.2.5 Electrowetting on Dielectrics

In addition to systems using physical structures for liquid handling, platforms based on "virtual channel structures" have been suggested, most notably in the form of digital microfluidics. These systems make up of the fact that contact angles of fluids on dielectric layers can be manipulated by applying electrical potentials to electrodes located underneath the dielectric layer. The correct reference term for these systems is "electrowetting on dielectrics (EWOD)" devices and the first occurrence in the literature dates back to 2002 (Lee et al. 2002). As the peripheral instrumentation required for operating these systems is rather ample, they are mostly limited to laboratory-scale experiments. However, the question of whether or not digital microfluidics is a suitable platform for LoC has been reviewed (Fair 2007). Further details about these systems can be found in recent reviews which the reader may refer to (Choi et al. 2012).

5.2.2.6 Centrifugal Microfluidics

Another commonly used platform for LoC devices is based on centrifugal forces. In these systems, microfluidic systems are designed in a disk-like structure similar to a compact disk (CD) with a central orifice for mounting onto a rotating axis (see Fig. 5.4).

The channels are arranged such that microfluidic flow can be modulated by varying the rotation speed of the axis. Doing so, one can set up microfluidic systems which allow fluid manipulation without using actively controllable components such as pumps and valves.

The first occurrence of this type of analytical platforms dates back to 1972 when Anderson and coworkers of Oak Ridge National Labs (ORNL) published their seminal paper on the development of the "Miniature Fast Analyzer" which described the centrifugal platform for the first time (Burtis et al. 1972). In 1989 Abaxis Inc. was founded, a spinoff which bought most of the ORNL patents on the Fast Analyzer with the aim of developing the centrifugal platform for whole blood assays. In 1998 Madou and coworkers introduced the LabCD as an analytical platform based on centrifugal actuation (Madou and Kellogg 1998). The paper introduced numerous fluidic concepts such as pumping, mixing, metering, and valving on the centrifugal platform and paved the way for establishing centrifugal microfluidics as suitable sample-to-answer type assay platform. In comparison to previous systems such as the Fast Analyzer, the LabCD can be operated without trained personnel as the microfluidic components and networks are designed such that a simple actuation protocol (i.e., spinning speed, spinning direction, and spinning duration of the axis) suffice for performing entire assays. The centrifugal platform poses some distinct requirements to component design. As the rotating

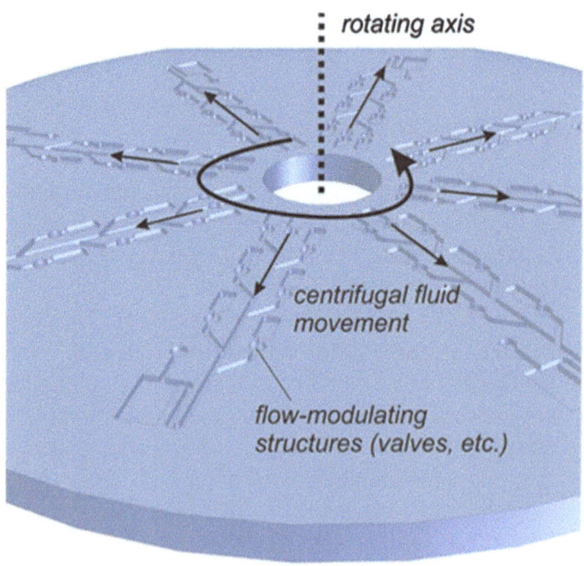

Fig. 5.4 Concept of centrifugal microfluidics. The main driving forces in these systems are the centrifugal forces exerted onto liquid inside of microfluidic channels if the channel substrate (i.e., the microfluidic chip) is rotationally accelerated. The flow is modulated by structures such as congestions which stop the flow unless a certain acceleration threshold is surpassed. This allows controlled movement of liquids in the microfluidic chip

disks are difficult to contact mechanically (and if so only temporarily), readout is mostly limited to optical means and active components (such as actively operated valves and pumps) are difficult to set up. Therefore centrifugal platforms are often operated using siphon valves, capillary valves, or hydrophobic barrier valves all of which are triggered by modulation of the rotation speed. Pumping is usually performed using centrifugal forces alone.

Although early commercial interest (such as the purchase of US startup company Gamera by Tecan) has spurred the development of the platform, most industrial activities have been suspended by the end of the 1990s, most notably Tecan's activities in the drug development based on the Gamera technology. Since 2000, centrifugal microfluidics has attracted significant academic interest and is among the most intensively student LoC platform. Today, numerous complex assays have been demonstrated on this platform such as an enzyme linked immunosorbent assay (ELISA) for the detection of Hepatitis from whole blood samples by Samsung (Lee et al. 2009) and several commercially available systems based on the Piccolo platform by Abaxis Inc. for kidney and liver function testing. These systems are also commonly used in veterinary diagnostics (Greenacre et al. 2008). For more information on the centrifugal platform, the reader may refer to the literature (Gorkin et al. 2010).

5.2.3 *Material Choice for PoC Devices*

Depending on the type of platform chosen, the material may be fixed. In case of capillary flow platforms, the material usually is cellulose or one of its common modifications (cellulose acetate or nitrate). The EWOD platform, relying on virtual channels, requires a dielectric substrate, usually with a hydrophobic fluorinated coating. Centrifugal microfluidics, especially if optical readouts are to be implemented relies on transparent materials of sufficient stiffness, and therefore, for the most part, on thermoplastic polymers. LoC devices on pressure-driven platforms are relatively tolerable to the choice of material which is mainly defined by the manufacturing technology chosen. Silicon is usually etched isotropically (e.g., using aqueous solutions of hydrazine or potassium hydroxide) resulting in semicircular channel cross sections. Similarly, glass can be etched isotropically using solutions of hydrofluoric acid (HF) (Fletcher et al. 2002). Plasma or ion beam etching (e.g., using reactive ion etching, RIE) allows anisotropic etching resulting in essentially rectangular channel cross sections. Details of these processes can be found in the literature (Delapierre 1989).

During late 1990s, microfluidics as a field emancipated from the classical MEMS structuring techniques, and more industrially accepted manufacturing methods such as polymer replication by means of hot embossing, thermoforming, or injection molding have been established (Heckele and Schomburg 2004). These techniques allowed creation of microfluidic structures at significantly reduced costs and on an industrial scale (Truckenmüller et al. 2002). One of the most important advantages of industrial polymer replication is the fact that these processes can be carried out outside of a clean room environment and require (depending on the feature resolution) significantly less elaborated machinery. This opened up the field of microfluidics and PoC to laboratories without access to clean rooms and thus the community was complemented by researchers more interested in the application of the structures rather than in their manufacturing. Common examples of thermoplastic polymers employed are polymethylmethacrylate (PMMA), polystyrene (PS), polypropylene (PP), or cyclic olefin copolymers (COC) with the latter becoming more and more important for optical assays due to its superior clarity and reduced background fluorescence. Other high performance polymers (Worgull et al. 2011) such as fluorinated polymers have also become of interest (Hannig et al. 2010). In 1998 the Whitesides group introduced PDMS, a room temperature curing elastomer which could be used to replicate microstructures from structurally weak templates such as lithographically structured layers of photoresist (most commonly the negative photoresist SU-8). The manufacturing method was termed soft lithography and can be considered a major milestone in the development of microfluidics.

Compared to polymer replication, soft lithography has the advantage of being applicable without the need of any equipment for replication at all. If suitable replication masters are at hand, the process can be carried out fast, reliably, and conveniently at room temperature. This simplification in device manufacturing has opened up the microfluidic and LoC field to users not experienced with setting up or operating instruments commonly used for polymer replication and thus allowed researches with backgrounds in biology, chemistry, and medicine to join the LoC community. The importance of this development has been highlight numerous times in the scientific literature (Kim et al. 2008). An overview about common used materials, their advantages and disadvantages and some examples is given in Table 5.1.

Depending on the choice of materials, several manufacturing techniques may be applicable which range from fast prototyping (Waldbaur et al. 2013a) to established industrial processes (Heckele and Schomburg 2004). Even platforms fully compatible with industrial standards have been described (Waldbaur et al. 2013b). Potentially techniques for sealing physical structures after manufacturing may need to be applied for which numerous physical and chemical means have been described. One has to note that sealing becomes significantly more challenging if different materials are to be permanently connected (Wilhelm et al. 2013b). Several reviews can be found in the literature that summarize the most commonly used materials for microfluidic PoC devices and the manufacturing techniques most commonly applied (Waldbaur et al. 2011).

Table 5.1 Common materials for PoC devices

Material	Advantages	Disadvantages	References
Cellulose	Low-cost	Ineffective sample consumption and sample evaporation	(Cheng et al. 2010; Vella et al. 2012; Mirica et al. 2012)
	No need of external pumping (due to capillary action)	Some hydrophobic barriers not strong enough for samples with low surface tension	
	Disposable	Usually high limits of detection	
Silicon	Well-rounded and well-characterized material	Not gas permeable → not suitable for cell culturing	(Erickson and Wilding 1993; Jensen 2006; Petralia et al. 2013)
	Cost-effective	High stiffness → unsuitable for mechanically movable components	
	Chemical and temperature resistance	Expensive compared to plastic devices	
		Opaque to visible and UV light	
Glass	High chemical and temperature resistance	Not gas permeable → not suitable for cell culturing	(Vrouwe et al. 2004; Easley et al. 2006; Lin et al. 2012)
	Easy surface modification	High stiffness → unsuitable for mechanically movable components	
	Very good cell adhesion	Expensive manufacturing	
	Good optical transparency		
	No autofluorescence		
PMMA	Low autofluorescence	Low stability against organic solvents	(Hong et al. 2010b; Ju et al. 2012; Lin et al. 2012)
	Biocompatible		
	Good optical transparency		
	Chemical inert in neutral aqueous solutions		
	Resistant to hydrolysis		
PS	Most common material for in vitro cell-based studies	Low gas permeability	(Recknor et al. 2006; Pfleging et al. 2009; Young et al. 2011)
	Biocompatible		
	Low water vapor permeability		
	Very good cell adhesion		

(continued)

Table 5.1 (continued)

Material	Advantages	Disadvantages	References
COC	Good chemical resistance	Low gas permeability	(Pu et al. 2006; Hong et al. 2010a; Schumacher et al. 2012)
	Low water absorption	High price	
	Good optical transparency up to near UV range	Low chemical resistance to aliphatic solvents	
	Long-term stable surface treatments		
	Low autofluorescence		
	Biocompatible		
	Low water vapor permeability		
	Good cell adhesion		
PDMS	Lower autofluorescence than COC, PMMA, or PC (Piruska et al. 2005)	Soft material → deformable → low pressure resistance	(Tan et al. 2011; van Kan et al. 2012; Horade et al. 2012)
	High gas permeability	Extensive swelling in organic solvents	
	Optical transparent from UV to NIR	Surface treatment unstable over time	
	Biocompatible	Absorption of hydrophobic compounds	
		Leaching of uncrosslinked oligomers	
		Evaporation of solvents (water)	
PC	Biocompatible	Low stability against organic solvents	(Yang et al. 2002; Witek et al. 2006; Kim et al. 2009)
	Optical transparent	Susceptible to stress cracking	
	High heat resistance		
	High impact resistance		

5.2.4 Commonly Used Components of LoC Systems

As discussed, numerous microfluidic platforms have been described in the literature and are being established as suitable operation units for LoC assays. Depending on the platform chosen, the choice of functional building blocks varies. As an example, implementing a microfluidic valve on a capillary flow platform is inherently different than implementing a valve on a pressure-driven platform. The following section will summarize the most important most commonly used components of LoC systems.

5.2.4.1 Microvalves

If microfluidics is considered the fluidic equivalent of microelectronics, the microvalve can be considered the transistor. It is the basic component that allows fluid manipulation by selectively hindering or allowing fluid flow in a system. There are several classification schemes for microvalves, most notably the classification into two categories: active or passive valves (see Table 5.2). Active valves usually require external actuation by means of a suitable actuator and therefore allow user interference at any time during operation. This is especially suitable during development of a LoC assay where the incubation intervals and similar parameters may not yet be fully known.

Passive valves always embed their operation principle within their physical or chemical structure and are triggered once certain physical or chemical properties of the fluid (such as its internal pressure) are exceeded. The general scheme of an active and a passive valve are shown in Fig. 5.5.

All microvalves share some common design hurdles that need to be overcome. Given the small dimension of the channels in question, the degree of perfection of the fluidic sealing is of paramount importance. In the macroscopic world fluidic sealing can rarely be accomplished by merely designing two components to match perfectly in shape (commonly referred to as form closure). One rather has to design the systems such that suitable sealing elements, e.g., in form of sealing rings or elastic components are sufficiently pressurized against a solid valve body in order to obtain sufficient sealing (commonly referred to as force closure). Due to the small dimensions of most microvalves, form closure is difficult to accomplish and sealing elements need to be manufactured at sufficiently small length scales in order to be applicable to the systems. Very often, fluidic sealing is facilitated by using rotation symmetric, self-centering and thus self-aligning structures such as balls or spheres. These are usually pressed against a suitable valve seat which is manufactured in form of a circular hole (Yamahata et al. 2005). These so-called ball valves allow bypassing the form/force closure problem commonly encountered when working with small dimensions. Another alternative is usage of elastic sealing elements which are actuated by means of non-directional force such as pneumatic or hydraulic pressure. Typically these sealing elements are designed in form of elastic membranes which deflect upon pneumatic pressure application and close a microfluidic channel.

Active microvalves are the most commonly used form of valves in laboratory setups as they allow opening and closing of channel structures upon user request. However, active microvalves always require some sort of actuation which usually also implies a suitable power supply and/or storage (such as in the form of a battery). Depending on the application scope of the LoC system to be designed, it may be necessary to limit the degree of fluidic complexity to as few energy-intensive components as possible. Reducing the complexity of the fluid handling on-chip usually allows reduction of the number of valves required. Several actuation principles for active microvalves have been described in the literature, some of

Table 5.2 Classification of microvalves

Principle	Operating mode	Actuation mechanism	Switching principle	References
Active	Mechanical	Magnetic	External magnetic fields	(Terry et al. 1979; Meckes et al. 1999; Bae et al. 2003; Rapp et al. 2010)
			Integrated magnetic inductors	(Sadler et al. 1999; Cho et al. 2001; Choi et al. 2001)
		Electric	Electrostatic	(Shikida et al. 1994; Goll et al. 1997; van der Wijngaart et al. 2002; Teymoori and Abbaspour-Sani 2005)
			Electrokinetic	(Kirby et al. 2002)
		Piezoelectric		(Peirs et al. 2000; Roberts et al. 2003; Waibel et al. 2003; Rogge et al. 2004)
		Thermal	Bimetallic	(Easley et al. 2006; Lin et al. 2012)
			Thermopneumatic	(Xing et al. 1999; Rich and Wise 2003; Takao et al. 2005; Yang et al. 2010)
			Shape memory alloy	(Kahn et al. 1998; Kohl et al. 1999b, 2000)
	Non-mechanical	Electrochemical		(Neagu et al. 1996; Hua et al. 2002; Suzuki and Yoneyama 2003)
		Phase change	Expansion	(Beebe et al. 2000; Carlen and Mastrangelo 2002; Klintberg et al. 2002; Baldi et al. 2003; Richter et al. 2003)
			Shifting	(Pal et al. 2004; Liu et al. 2004b; Yang and Lin 2009; Ogden et al. 2010)
			Freeze-thaw	(Gui and Liu 2004; Chen et al. 2005; Gui et al. 2011; Neumann et al. 2013)
		Rheological	Electro-rheological	(Yoshida et al. 2002)
			Ferrofluids	(Hartshorne et al. 2004)
	External	Pneumatic	Membrane	(Kazuo and Ryutaro 2000; Grover et al. 2003; Go and Shoji 2004; Luque et al. 2005)
			In-line	(Unger et al. 2000; Studer et al. 2004; Wang et al. 2004)

(continued)

Table 5.2 (continued)

Principle	Operating mode	Actuation mechanism	Switching principle	References
Passive	Mechanical	Check valve	Flap	(Zengerle and Richter 1994; Koch et al. 1997; Kwang-Seok et al. 2002)
			Membrane	(Bien et al. 2003; Chung et al. 2003; Li et al. 2005)
			Ball	(Carrozza et al. 1995; Yamahata et al. 2005)
	Non-mechanical		Diffuser	(Gerlach 1998; Olsson et al. 2000; Schabmueller et al. 2002)
		Capillary	Abrupt	(Ahn et al. 2004)
			Liquid triggered	(Melin et al. 2004b)
			Burst	(Duffy et al. 1999; Johnson et al. 2001; Puckett et al. 2004)
			Hydrophobic valve	(Andersson et al. 2001a, 2001c)

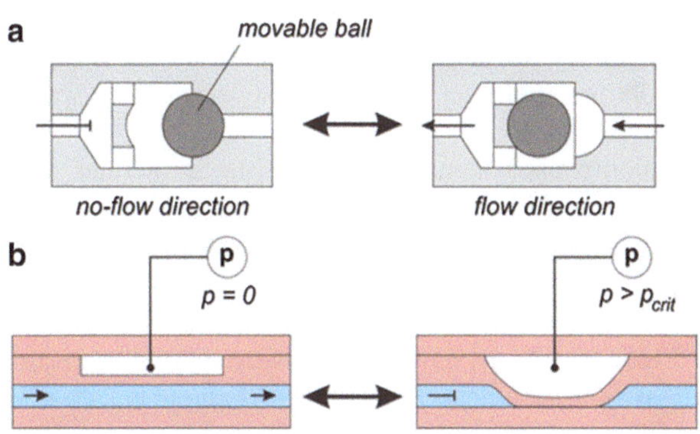

Fig. 5.5 Microfluidic valve structures. (**a**) Passive one-way structure using a physically movable ball. Depending on the flow direction, the ball will be pressed onto a valve seat or away from it. This allows for flow rectification. (**b**) Active microfluidic valve structure at the example of a pneumatically actuated membrane valve. An externally controlled pressure source is applied to the top channel structure which is commonly referred to as control chamber. If under pressure, the elastic separating membrane is deflected and seals off the fluidic channel. These valves are commonly used in microfluidic systems which are entirely manufactured from soft elastomers such as PDMS

which lend themselves very well to small, portable LoC and some of which are predominantly limited to lab applications where constant power supply is available.

Magnetic actuation was the first operation concept of a microfluidic valve which dates back to the work of Terry et al. (Terry et al. 1979). Two versions of these valves are commonly described in the literature. The first design places the magnet and the entire actuation outside of the valve. Essentially, the valve is then actuated by a linear movement exerted by an external actuator. Examples include designs where flexible membranes are actuated (Byunghoon et al. 2003) as well as ball-type valves (Fu et al. 2003). The second design uses a soft magnet such as a nickel/iron alloy, embedded into the actuation membrane. Applying a magnetic field will actuate the membrane thus closing the valve (Sadler et al. 1999; Choi et al. 2001).

Among the oldest and most commonly described actuation concept for LoC microvalves is piezoelectric actuation. The first report dates back to 1975 (Thomas and Bessman 1975) with first applications as insulin pump being published as early as 1978 (Spencer et al. 1978). These first systems were usually set up non-monolithically and the first all-silicon-based components were demonstrated by Terry et al. (Terry et al. 1979) and developed further by Smits (Smits 1985, 1990) and van Lintel et al. (van Lintel et al. 1988). Piezoelectricity refers to the effect that certain crystals (due to the symmetry of their charge distribution) will influence charge separation along their axis if mechanically deformed. The inverse effect, applying an electric potential which results in a deformation, has been used as actuation principle for a very long time. Several commercially available microvalves and micropumps are based on this concept. These components can be obtained e.g., from Bartels Mikrotechnik (Germany), thinXXS (Germany), or Star Micronics (Japan). Numerous examples of microvalves based on piezoelectric actuation can be found in the literature with emphasis on the development of valves with low dead volumes (Peige et al. 2004), fast reactions or specific application scenarios (Chakraborty et al. 2000).

Another commonly used actuation mechanism is thermal actuation. Ohmic heating can be implemented conveniently and lends itself well for designing LoC systems and assays where high-density arrays of valves are required. In general, thermal expansion alone is not sufficient to create sufficient strokes for valve actuation, but the bimetallic effect can be exploited for doing so (Barth 1995; Jerman 1994). Phase transitions have been another option for creating large-stroke valves. Liquid/gas transitions can be exploited (Fahrenberg et al. 1995; Bocong et al. 2010) as well as solid/liquid transitions, especially when aliphatic hydrocarbons or carboxylic acids are used which undergo structural rearrangements upon melting resulting in significant deformation. Large-scale valve platforms have been set up using this concept (Neumann et al. 2013) as well as ultrafast valves (Feng and Chou 2011). As the manufacturing of these valves usually requires non-hybrid setups, attempts have been made to design the valves as surface-machined planar structures (Carlen and Mastrangelo 2002). This type of valve can also be used as single-use component. In such structures a channel is opened when the phase-change material (usually a wax) is molten. This effect cannot be reversed: Once molten the valve cannot close again (Liu et al. 2004b). This type of disposable valve is very cheap and thus suitable for PoC diagnostics. Shape memory alloys have also been commonly used as actuator materials for setting up microvalves (Kohl et al. 2000, 1999a). Upon application of a magnetic field and temperature, these materials undergo changes in their phase morphology which results in significant changes in their lateral dimension.

As stated, electrokinetics has been a convenient mechanism for fluid actuation in microfluidics which is predominantly used for pumping. However, electrokinetic pumping can also be used in order to set up microvalves, e.g., if an in situ polymerized polymer plug is displaced against a valve seat using an electrokinetic pump (Kirby et al. 2002). Electrostatics can also be used as actuation principle. Here, a charged electrode will attract an inversely charged actuator surface (usually embedded in the sealing membrane) to close the valve temporarily. Upon removal of the electric field, the elastic membrane will relax thus opening the valve. Nguyen et al. manufactured an electrostatically actuated valve using an ionic polymer/metal composite. This composite was used to create the actuator membrane which could be deflected by application of up to 3 V (Nguyen et al. 2008). This effect can also be employed to stabilize the open- or closed-state of a value actuated by different mechanisms (Wagner et al. 1996).

Another commonly described class of microvalves is based on stimuli responsive materials which are sometimes referred to as smart materials. These actuators respond to changes in the ambient conditions, such as pH changes or changes in the composition of the fluid. Potential responses include extensive swelling and thus blocking of channels. Liu et al. have demonstrated stimuli-responsive microfluidic valves by structured photopatterning of photocurable hydrogels inside of microfluidic channels. The hydrogel, which was lightly cross-linked poly (methacrylic acid), is a solid polymer at pH values below 7 and deprotonates to give a highly swellable absorber above pH 10. Under such liquid flows, the microfluidic channels closed due to swelling of the valve thus blocking the flow (Liu et al. 2002). It was also shown, that such valves can also be actuated externally, thus avoiding the necessity of probing the swelling solution through the primary channel network. In these systems, the valve is arranged such, that a second channel connects to the hydrogel structure which may be probed and thus swollen using this second channel (Beebe et al. 2000). Embedding the actuator material in a separate chamber which is sealed with a valve membrane and which (upon expansion) closes the microfluidic channel is another potential concept (Liu et al. 2002).

Another mechanism based on electrical fields is usage of electro-rheological fluids. These fluids are usually made up of an electrical insulator (such as paraffin oil) containing particles that undergo polarization in an electric field and thus orient along the field lines. As a consequence, the material stiffness and the flow is inhibited. This type of actuation has also been described for nematic liquid crystals (Yoshida et al. 2002). Similar effects can be obtained when using ferrofluids, which are liquids with suspended magnetic particles which undergo similar reorientation and viscosity changes in magnetic fields (Hartshorne et al. 2004). However, electrorheological as well as ferrorheological effects usually require high electric and magnetic field strengths which makes them unsuitable for high-density integration of microvalves.

Pneumatic actuation is among the most important actuations principles for microfluidic valves since the introduction of PDMS as elastomeric bulk material for microfluidics (Duffy et al. 1998). In 2000 the Quake group introduced monolithic microvalves based on the elastic properties of PDMS. These so-called NanoFlex valves are designed monolithically and feature microfluidic channels on two levels, separated by a thin membrane (Unger et al. 2000). The upper layer (the so-called control layer) features channels which can be pressurized. In pressurized state, the thin membranes will deflect and seal the channels in the lower layer resulting in a microvalve. Several valves can be operated by the same control channel which makes the development of large-scale systems with hundreds to thousands of mechanical valves possible (Thorsen et al. 2002). This technique is generally referred to as microfluidic large-scale integration and has allowed setting up LoC systems with unmatched complexity. The main disadvantage of this concept is given by the solvent compatibility of PDMS which limits the application scope of these systems to aqueous solutions. Interestingly, the concept of using PDMS membranes bonded to a structured layer pressure guiding structures resulting in a pneumatic module was already developed before (Kazuo and Ryutaro 2000) as was the usage of elastic materials such as latex membranes (Lagally et al. 2000) and even PDMS membranes (Ohori et al. 1998; Yang et al. 1998). These concepts had already been extended to the design of multi-way membrane valves at the time (Kazuo and Ryutaro 2000).

In addition to these integrated microvalves, several concepts have been described in the literature that provide external valving and merely interface the external valves with the microfluidic structures of the LoC system. Examples include the adaptation of rotary valves which are commonly used in liquid chromatography (Hasegawa et al. 2008) and selected chip-to-world interfaces that allow valving (Oh et al. 2005).

Passive microvalves are valves that do not require a certain stimulus in order to trigger. A typical example is a check valve which allows fluid movement in one direction, but not in the other. Passive vales must be designed such that they fulfill their purpose robustly by design, as user interference (or control) as such is not possible. However, the passive nature of their structure has a number of advantages. They are mostly easy to design and robust in operation. As passive valves do not require an actuator, there is no need for external power supply, which makes them extremely suitable for decentralized applications in PoC diagnostics.

As stated, check valves are the classical example of passive microvalves. They allow flow only in one dimension and are therefore sometimes referred to as or used for microfluidic diodes (Adams et al. 2005). In essence, two types of check valves are known from the literature. Flap valves, which make use of elastic membranes that bend according to the flow direction. Depending on the flow direction, the flap can be deflected by the liquid such that it either clears or blocks the fluid passage. The second commonly used check valve is the so-called diaphragm valve. This valve operates similarly to the flap valve but uses a perforated diaphragm sandwiched between two channels. Depending on the flow direction, the liquid either transits the diaphragm through the orifice or presses the diaphragm against one of the adjacent channel thus blocking the flow. Both concepts have been thoroughly employed in microfluidics (Jeon et al. 2002).

An alternative to flap and diaphragm valves are check valves based on mechanically freely movable objects such as beads, balls, or spheres. If these objects are enclosed in a microfluidic chamber with suitable formed outlet structures, the fluid flow, if directed correctly, can push the object against the outlet structure thus blocking the flow (Yamahata et al. 2005). One has to note, that embedding freely movable objects in microfluidic systems, which are mostly created using two-dimensional manufacturing techniques, is not a trivial task and significantly increases manufacturing complexity and costs. Here, an approach based on the in situ polymerization of a movable object, in form of a highly fluorinated polymer plug, have been described and successfully applied to the design of microfluidic valves (Rehm et al. 2001). Jeon et al. describe an all-polymer microfluidic pump and valve setup which is manufactured entirely from PDMS. The authors describe the design of passive flap and diaphragm type valves (Jeon et al. 2002).

Selective-stop valves are valves that allow selective flow of liquids, depending on their physical or chemical property. Most commonly used (also in microfluidic systems) are valves that allow air flow but prevent liquid flow. These so-called vents are often used to allow air bubbles to escape a microfluidic channel network. They are commonly employed when liquids are transferred onto a LoC system which can rarely be accomplished without unintentionally also probing air into the system. In its simplest form, such valves are set up in form of hydrophobic membranes (such as porous Teflon® membranes). They were first described by Anderson et al. in 1997 (Anderson et al. 1997) but can be found in numerous LoC systems known from literature (Lagally et al. 2000, 2001; Belgrader et al. 2000).

The micro regime lends itself well for tailoring the contact angles of surfaces inside of channels by suitable surface modifications. Small changes in the surface chemistry can lead to tremendous changes in the wetting behavior. Zhao et al. used laminar flows of silanes and photodeprotection of nitrobenzene-functional silanes to create channels with selective hydrophilic/hydrophobic wetting behavior (Zhao et al. 2001). Selective channel patterning can not only be used for hindering fluid flow but also for the creation of pressure barriers. If the fluid pressure increases above a certain threshold, the fluid breaks out of the hydrophilic surface areas and wets the hydrophobic channel sections as well. This concept is equivalent to a burst valve and may lend itself well for programming fluid logics or assays.

Burst valves are a type of valve commonly found in assays that require on-chip storage of reagents. These reagents are often encapsulated in suitable chambers, vials or tubes which need to be broken in order to free the liquid for the assay. Besides mechanical means, similar concepts can be set up in small microfluidic channels by exploiting the retention forces of a liquid meniscus due to surface tension. Such menisci are stable up to a certain back pressure whereupon they break to allow liquid flow. Increasing backpressure is convenient in some platforms, such as centrifugal microfluidics where these types of valves are commonly used to program assays (Zeng et al. 2000). Adjustment of surface wetting has been described for these systems as well, which allows the bursting properties (i.e., the required backpressure) to be adjustable over a wide range (Tiensuu et al. 2000).

Further details about the various types of microvalves can be found in the excellent review by Oh and Ahn to which we refer the reader for further information (Oh and Ahn 2006).

5.2.4.2 Micropumps

With respect to their design, micropumps share a number of design concepts with microvalves. A classification of different micropumps is shown in Table 5.3. This is due to the fact that most microvalves, upon closure, displace a certain volume of fluid due to the displacement of the sealing element. Once the valve actuation is switched off, the valve resets to its original state, potentially moving the displaced volume again. If this displacement/suction motion can be directed, a pump can be designed. Usually three displacement valves aligned in a row make up a micropump. Typical examples include elastomeric membrane valves in PDMS (Chou et al. 2001; Unger et al. 2000). Often, micropumps are set up using one actively operated pump valve and two check valves for directing flow movement. Sin et al. describe such a micropump which consists of a pneumatically actuated elastomer membrane for pumping and two passive ball check valves (Sin et al. 2004). Yamahata et al. describe the combination of a magnetically actuated microvalve acting as a driving valve in combination with two passive ball-type check valves (Yamahata et al. 2005). Nguyen et al. demonstrated an electrostatically actuated valve using an ionic polymer/metal composite as the driving valve (Nguyen et al. 2008). Jeon et al. developed all-polymer microfluidic systems made from PDMS demonstrating how flap and diaphragm valves can be set up monolithically in soft elastomers. The authors also show, how a simple microfluidic pump, actuated manually, can be created by the combination of two check valves and one membrane valve which is pumped by hand (Jeon et al. 2002). Fu et al. demonstrated an electrically actuated micropump using a ball valve that was actuated by a commercially available coil (Fu et al. 2003). Wego et al. demonstrated a pump using two passive valves and one actively actuated pumping valve using thermopneumatic actuation (Wego and Pagel 2001). Böhm et al. demonstrate electromagnetic actuation using the same principle setup (Böhm et al. 1999b).

Khoo and Liu use magnetic actuation with a soft magnet integrated into the sealing membrane (Khoo and Liu 2001). Benard et al. demonstrate usage of shape memory alloys as actuation element (Benard et al. 1998).

Some micropumps can be set up without additional flow directing valves by using specifically shaped inlet and outlet channels resulting in diffuser structures. One example of such a design is presented by Andersson et al. using a piezoelectric actuator (Andersson et al. 2001b). Electrolyte solutions can be pumped using Lorentz force based actuation (Lemoff et al. 1999). Inline micropumps can also be set up using a second fluid as driving liquid such as a ferrofluid. Using a rotating magnet, the ferrofluid plugs are moved along the channel dragging the bulk liquid along (Hatch et al. 2001). In addition, almost any effect that creates bulk liquid

Table 5.3 Classification of micropumps

Principle	Operating mode	Stimulation	References
Displacement	Reciprocating—diaphragm	Piezoelectric (lateral, axial)	(van Lintel et al. 1988; Smits 1990; Gass et al. 1994; Olsson et al. 1995; Gerlach and Wurmus 1995; Nguyen and Huang 2001; Gu et al. 2004; Johnston et al. 2005; Esashi et al. 1989; Li et al. 2000)
		Thermopneumatic	(Van de Pol et al. 1990; Mizoguchi et al. 1992; Folta et al. 1992; Jeong and Yang 2000; Wego and Pagel 2001; Tsai and Liwei 2002; Sim et al. 2003)
		Electrostatic	(Judy et al. 1991; Zengerle et al. 1992, 1995)
		Pneumatic	(Rapp et al. 1994; Unger et al. 2000; Chou et al. 2001; Grover et al. 2003; Sundararajan et al. 2005; Kim et al. 2006; Lien et al. 2007; Huang et al. 2008; Jeong and Konishi 2008)
		Shape-memory Alloy	(Benard et al. 1998)
		Electromagnetic	(Dario et al. 1996; Böhm et al. 1999b)
		Electrowetting	(Kwang-Seok et al. 2002)
	Reciprocating—piston	Magnetic	(Münchow et al. 2005)
	Rotary		(Ahn and Allen 1995; Sen et al. 1996; Döpper et al. 1997; Hatch et al. 2001; Terray et al. 2002)
	Aperiodic	Pneumatic	(Jen and Lin 2002; Tas et al. 2002)
		Phase change (thermal, electrochemical)	(Takagi et al. 1994; Jun 1998; Lee et al. 1998; Böhm et al. 1999a; Handique et al. 2001; Lui et al. 2010)
		Electrowetting	(Colgate and Matsumoto 1990; Lemoff and Lee 2000; Lee et al. 2002; Pollack et al. 2002)
		Thermocapillary	(Burns et al. 1996; Sammarco and Burns 1999; Kataoka and Troian 1999)
		Osmotic	(Yardley and Linkenheimer 1971; Su and Lin 2004)

(continued)

Table 5.3 (continued)

Principle	Operating mode	Stimulation	References
Dynamic	Centrifugal		(Marseille et al. 1998; Duffy et al. 1999; Fréchette et al. 2000; London et al. 2001)
	Electrohydrodynamic	Injection	(Richter and Sandmaier 1990; Richter et al. 1991; Ahn and Kim 1997; Darabi et al. 2002)
		Induction	(Bart et al. 1990; Fuhr et al. 1992, 1994; Fuhr 1997)
		Conduction	(Jeong and Seyed-Yagoobi 2002; Atten and Seyed-Yagoobi 2003)
	Electroosmotic		(Dasgupta and Liu 1994; Ramsey and Ramsey 1997; McKnight et al. 2001; Laser et al. 2002; Chen and Santiago 2002; Zeng et al. 2002)
	Magnetohydrodynamic	DC	(Jang and Lee 2000)
		AC	(Lemoff and Lee 2000)
	Acoustic streaming/ ultrasonic		(Moroney et al. 1991; Kurosawa et al. 1995; Nguyen et al. 2000; Rife et al. 2000; Guttenberg et al. 2005; Tovar and Lee 2009)
	Capillary		(Juncker et al. 2002; Jensen 1998; Knight et al. 1998)

movement can be used as a pump. Here typical examples include, among others, electro-rheological effects (Liu et al. 2006) and pumps based on electroosmotic flow (Zeng et al. 2001; Chuan-Hua and Santiago 2002).

Interestingly, a number of passive pumping concepts have been suggested over the years. These pumps do not require any movable mechanical components at all and are therefore very robust in operation and simple in setup. One example of such a pump was described by Geng et al., using expanding and collapsing bubbles in order to drive flow (Geng et al. 2001). Osmotic pumps have also been suggested although their pump rates are relatively small (Yu-Chuan et al. 2002). Another important class of pumping principles includes passive capillary pumping, first described by the Beebe group in 2002 (Walker and Beebe 2002). Passive pumping makes use of the fact that if two liquid droplets exposed to air can communicate via a microfluidic channel, the smaller droplet will pump liquid into the bigger droplet. This is due to the fact that smaller droplets have higher surface tensions due to their higher curvature (Berthier and Beebe 2007; Chen et al. 2009). A number of assays have been set up using this passive pump platform (Meyvantsson et al. 2008). Examples of pumping structures are shown in Fig. 5.6.

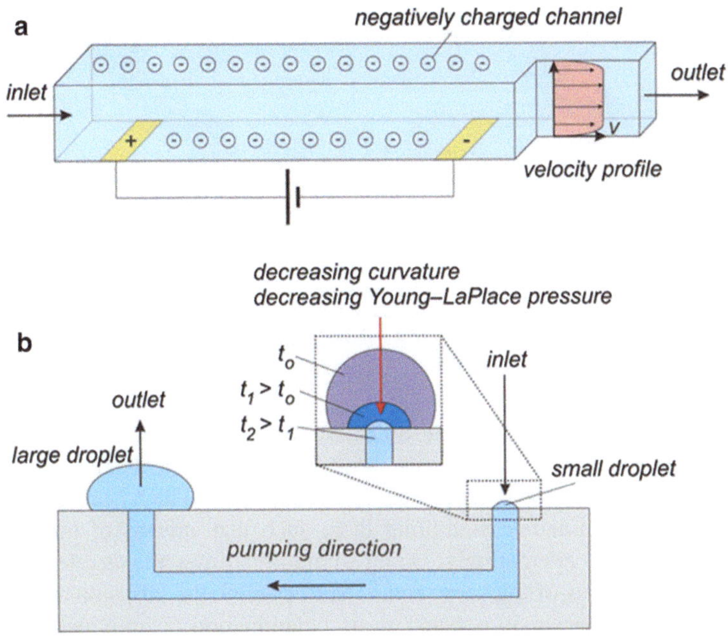

Fig. 5.6 Examples of typical microfluidic pumping structures. (**a**) DC electroosmotic flow at the example of a negatively charged channel. This is the case, e.g., for fused silica glass tubes at pH values above 4. The deprotonated silanol groups at the surface will result in a negative net charge. Applying a potential across the channel, e.g., by means of two electrodes, results in a net flow due to the drag of the electroosmotic flow (EOF). (**b**) Passive pumping structure used with a microchannel to which a small and a larger droplet are applied. Due to the difference in droplet curvature (and therefore in different Young-LaPlace pressure), the smaller droplet will be passively pumped into the larger droplet

5.2.4.3 Micromixers

Given the small feature sizes of fluidic structures in most LoC, the fluid flow in these systems is strictly laminar. This can be quantified using the Reynolds number, a dimensionless quantity which represents the ratio of inertia to viscous forces. If this number is high, the flow is turbulent, if the number is low (usually below a threshold of 1,000–2,300, depending on the source cited) the flow is strictly laminar. Typical LoC systems show Reynolds numbers below 10. Laminar flow has a number of advantageous properties such as defined diffusion over short distances, which allows the generation of complex gradient patterns, e.g., for cell studies (Dertinger et al. 2001). However, this may also be considered an inherent disadvantage if thorough mixing is required. For this, several concepts of micromixers have been developed which allow effective mixing at low Reynolds numbers.

In general two categories of micromixers are known from literature: active mixers (which require some sort of external actuation) and passive mixers.

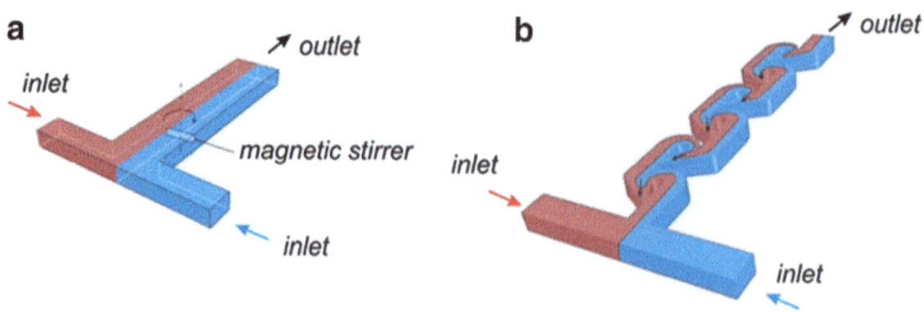

Fig. 5.7 Examples of mixing structures commonly used in microfluidics. (**a**) Magnetically stirred microchannel using a rotating external magnetic field in combination with an inline magnetic stirrer. (**b**) Modified in-plane Tesla mixing structure. This structure is commonly used to create multiple laminated flows and thus good diffusive mixing even at low Reynold numbers

The latter category lends itself well to LoC systems and has thus been extensively used. Examples of mixing structure are shown in Fig. 5.7. In general mixing in LoC systems usually implies splitting up two incoming flows (which need to be mixed) into smaller streams and reassembling them such that streams of the incoming streams are arranged alternatingly. Doing so multiple times allows effective diffusive mixing if streams of thickness equivalent or below the diffusion distance can be created. This concept is referred to as colamination or split-and-recombine mixing and can either be done in-plane (Bessoth et al. 1999; Koch et al. 1998) or out-of-plane (Hinsmann et al. 2001; Norbert et al. 1996). In-plane mixers are significantly easier to set up. One of the most effective in-plane mixers known is the so-called Tesla mixer which uses the Coandă effect, the tendency of a liquid to follow a convex surface rather than detaching and following the original flow line direction. This can be conveniently exploited to create streams with opposing flow directions and thus chaotic mixing which is difficult to obtain in the micro regime. The first occurrence of the Tesla mixer dates back to 2001 (Hong et al. 2001) with numerous improvements made to the original structure (Hong et al. 2004). Another in-plane mixer (which albeit requires two stacked planar structures) is the staggered herringbone mixer introduced in 2002 by the Whitesides group (Stroock et al. 2002). In addition to colamination or split-and-recombine mixing, the formation of water-in-water droplets (usually from two flows of the same nature, e.g., aqueous solutions) is also an effective way to reduce the diffusion distances and result in effective mixing (Tice et al. 2003).

Active micromixers can be set up if at least one of the flows can be actively moved, for example by ultrasound actuation (Yang et al. 2000), electrokinetic movement (Oddy et al. 2001), or mechanical stirring (Lu et al. 2001). Naturally, pumping is also a suitable means for mixing, especially if the pump is designed as a rotary system (Chou et al. 2001). Further details about commonly used micromixer concepts can be found in the literature (Nguyen and Wu 2005) and a classification is given in Table 5.4.

Table 5.4 Classification of micromixers

Principle	Actuation principle		References
Active (fluid displacement)	Pressure		(Deshmukh et al. 2000; Niu and Lee 2003; Glasgow and Aubry 2003; Fujii et al. 2003)
	Electrohydrodynamic		(El Moctar et al. 2003)
	Electrokinetic		(Oddy et al. 2001; Biddiss et al. 2004; Fu et al. 2005; Huang et al. 2006; Yan et al. 2009)
	Magnetic		(Bau et al. 2001; Lu et al. 2002; Suzuki and Ho 2002; Wang et al. 2008; Wen et al. 2009)
	Piezoelectric		(Woias et al. 2000)
	Thermal		(Tsai and Lin 2002; Legendre et al. 2006)
	Acoustic/Ultrasonic		(Tsao et al. 1991; Yang et al. 2000; Yasuda 2000; Yaralioglu et al. 2004; Ahmed et al. 2009)
	Stacking principle	Microfluidic structure	
Passive	Lamination	T-mixer/Y-mixer	(Bökenkamp et al. 1998; Veenstra et al. 1999; Gobby et al. 2001; Wong et al. 2004; Legendre et al. 2006)
		Multiple lamination	(Koch et al. 1999; Bessoth 1999; He et al. 2001; Hessel et al. 2003; Melin et al. 2004a)
		Split and Recombine	(Branebjerg et al. 1996; Schwesinger et al. 1996; Munson and Yager 2004; Schönfeld et al. 2004)
		Hydrodynamic focusing	(Jensen 1998; Knight et al. 1998; Veenstra et al. 1999; Walker et al. 2004)
	Injection		(Miyake et al. 1997; Song et al. 2003; Tice et al. 2004)
	Chaotic	Obstacles/Holes on wall (multilayer)	(Stroock et al. 2002; Kim et al. 2004; Kang and Kwon 2004)
		Embedded barriers (single layer)	(Wang et al. 2002; Hong et al. 2004; Singh et al. 2009; Tsai and Wu 2011)
		Meander-like, zig-zag or twisted channels	(Liu et al. 2000; Mengeaud et al. 2002; Jiang et al. 2004; Park et al. 2004)

5.2.4.4 Filters

Filtration ranges among the most commonly employed technique in LoC devices especially if crude sample matrices such as saliva or blood, need to be processed. The classical approach to filtration involves in-flow obstacles which retain components transported within the flow according to their size. Naturally, these techniques can be refined to all types of chromatographic separation methods which may even involve sample extraction, preprocessing or preconcentration. On chip filters have been created, e.g., using deep reactive ion etching (DRIE) (Andersson et al. 2000) or using ultrafiltration in flow-based separation (Andersson et al. 2001c). On centrifugal platforms, filtration and separation can also be carried out using centrifugal forces (Strohmeier et al. 2013). Often, filters are also set up using membranes, a technique for which detailed discussions can be found in the literature (de Jong et al. 2006). As discussed, solid phase extraction can also be an option for sample filtration. Here the sample of interest is temporarily bound to an adsorber after which the bulk of the liquid is flushed out of the system (Huh et al. 2007). One has to keep in mind that obstructive filtration always carries the risk of device clogging—this is especially problematic, if the channel dimensions of the system are fine and the sample matrix contains large portions of non-dissolved matter. In order to circumvent this problem, in-flow filtration using deterministic lateral displacement, first described in 2004 by Huang et al. (Huang et al. 2004), can be used (see Fig. 5.8). These systems make use of the fact that a flow field passing around an array of obstacles temporarily accelerates during bypassing. If particles are transported along the flow, these particles may not be able to keep up with the flow as they are transported out of the flow lines due to centrifugal forces. In such a

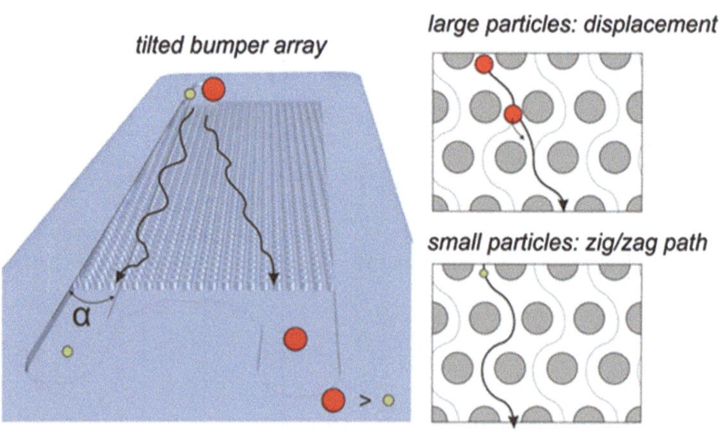

Fig. 5.8 In-flow filtration using deterministic lateral displacement. These filters are commonly referred to as microfluidic bumper arrays. The separation principle is based on the deviation of larger particles in a tilted array of microfluidic pillars. They are moved out of the shortest streamline path as they are too heavy and are dragged out of the streamlines due to centrifugal force. Smaller particles remain with the streamlines and thus downstream separation occurs

case, heavier and bulkier particles are continuously transported out of the flow field laterally and thus displaced in a deterministic way. These systems have proven to be suitable non-clogging filters if for example raw blood is used as sample matrix, e.g., for the extraction of circulating tumor cells (Nagrath et al. 2007).

Several other mechanism for in-flow filtration can be exploited which have been summarized in the literature (Pamme 2007).

5.2.4.5 Sensors

Various types of sensors have been described for LoC systems based on, among others, optics, electrochemistry, gravimetry, or acoustics. Different kinds of sensors used in LoC devices can be found in Table 5.5. In general, sensors for LoC systems are designed to either measure physicochemical or biochemical parameters. One of the most commonly, and most conveniently measured physical quantity is temperature. Temperature can be measured using thermoresponsive materials such as platinum resistors (e.g., PT100 or PT1000) which change their electrical resistance upon heating. Various other means can also be applied, e.g., thermoresponsive dyes (Ryu et al. 2009; Ross et al. 2001) which change their color upon heating.

Addition physical parameters measured include density and viscosity for which approaches based on suspended mechanical resonators (Khan et al. 2013) have been described. Flow velocity can be determined on the system level by metering the pumped liquid over time by inline flow sensors. Commonly used principles for doing so include heat transfer by the so-called anemometric sensors and visual tracking, e.g., using fluorescent dyes. One of the first flow sensors was described by Branebjerg et al. in 1991 (Branebjerg et al. 1991). The authors used the anemometric principle for detecting flow velocity as a function of heat transfer from a heater to a detector. Another principle described early is based on measuring the pressure drop of a fluid which can be correlated to the flow speed (Cho and Wise 1993). Early reviews made during the early 1990s still summarize the most relevant principles for measuring liquid flow inline and online (van Oudheusden 1992). In electroosmotic flows, the current can also be measured as means of determining the flow velocity (Huang et al. 1988). Sometimes, the measurement of the flow profile alongside the flow speed may be required which can be conveniently performed using particle tracking (Santiago et al. 1998). Conductivity is usually measured electrically by monitoring the change in the capacitance of an electrode embedded in the microfluidic channel. Alongside the conductivity, the change in the dielectric properties of the fluid may be required which, in terms, allows monitoring the change of surface impedance. pH sensing is another frequently employed technique for characterizing physicochemical changes in the liquid (Lin et al. 2006).

LoC systems are currently predominantly employed for detecting biochemical changes inside the channels. The applications demonstrated using LoC technology have ever increased since the manufacturing technology has matured sufficiently to

Table 5.5 Different Sensors used in PoC/LoC devices

Sensing mechanism		Transduction mechanism	Quantity to be measured (exemplary)
Electrochemical	Amperometry	Measurement of current in oxidation or reduction process	Detection of different pathogens (e.g., *Bacillus anthracis*, *Escherichia coli*, and influenza A virus) (Ghindilis et al. 2007)
			Respiratory activity of a biofilm (Pires et al. 2013)
			Concentration of blood glucose (Pickup et al. 1989)
	Impedance	Measurement of impedance of the system as a function of excitation frequency	Dielectric characterization of a cell population (Cheung et al. 2005)
			Detection of presence of a bacterial film on electrode surface (Pires et al. 2013)
	Potentiometry	Measurement of potential changes usually translated in a voltage change	Amount of DNA molecules on sensor (Barbaro et al. 2006)
			Detection of target DNA sequences (Star et al. 2006)
			Examination of pH value (Lin et al. 2006)
	Voltammetry	Measurement of current as a function of potential	Detection of pathogens (*E. coli*) and identification of genotypes (HFE-C gene) (Liu et al. 2004c)
			Concentration of free electroactive deosynuleoside triphosphates (dNTPs) for indirect monitoring of amplified DNA during PCR (Deféver et al. 2009)
Gravimetric	SAW	Measurement of frequency change (change of acoustic velocity) as indication of adsorbed mass on sensor surface	Detection of hepatocyte growth factor in cell culture medium (Berger et al. 2010)
			Detection of chemical agents (e.g., acetonitrile and dichloromethane) (Joo et al. 2007)
			Detection of penicillin G in milk (Gruhl and Länge 2014)
	QCM		Determination of albumin concentration (Lin et al. 2004)
			Detection of *Bacillus anthracis* (Hao et al. 2009)

(continued)

Table 5.5 (continued)

Sensing mechanism		Transduction mechanism	Quantity to be measured (exemplary)
Optical	Fluorescence	Measurement of fluorescence	Concentration of oxygen during cell culture experiments (Mehta et al. 2007)
			Concentration of green alga *Chlamydomonas reinhardtii* for water pollutant detection (Lefèvre et al. 2012)
	Absorbance	Measurement of absorbance (as a function of wavelength)	Concentration of uric acid in urine samples (Minas et al. 2004)
			Detection of caffeine, paracetamol, ketoprofen, and ascorbic acid during electrophoretic separation (Petersen et al. 2002)
	Luminescence	Measurement of Luminescence	Concentration of hydrogen peroxide produced while oxidation of glucose or ethanol released from immobilized yeast cells (Davidsson et al. 2004)
			Detection of ATP as indication of cellular metabolism (Liu et al. 2004a)
	Interferometry	Measurement of phase change by means of refractive index change	Detection of HSV-1 virus in serum (Ymeti et al. 2006)
			Detection of streptavidin (Hong et al. 2006)
	Colorimetry	Digital color analysis	Determination of pH and HSA and glucose concentration (Abe et al. 2008)
			Determination of pH and IgG concentration (Abe et al. 2010)

(continued)

Table 5.5 (continued)

Sensing mechanism		Transduction mechanism	Quantity to be measured (exemplary)
Electrical	Resistance	Measurement of resistance change	Temperature control of microfluidic chamber (Lao et al. 2000)
			Pressure sensing within PDMS microfluidic systems (Wu et al. 2011)
			Determination of flow rate in microfluidic channels (Rasmussen and Zaghloul 1999)
	Capacitance	Measurement of capacity change	Determination of position and volume of droplets (Chen et al. 2004)
			Detection of organic vapors (e.g., toluene and isopropyl alcohol) (Satyanarayana et al. 2006)
	Inductance	Measurement of change of inductive reactance	Detection of target molecules (DNA samples) (Wang et al. 2009)
			Detection of metallic debris in nonconductive lubrication oil (Du and Zhe 2011)

allow researcher to focus predominantly on the application of the systems rather than on engineering of components. As stated, the development of polymer based LoC devices, especially systems made from PDMS, have supported this development greatly. Current research not only focusses on biomedical assay development but also on the development of tools for large-scale and highly parallel biological screening such as single cell analytics which has become one of the major application fields of LoC devices recently (Kortmann et al. 2011). LoC systems are also being applied as tools in environmental monitoring or for the characterization of, e.g., bacterial biofouling on technical surfaces (Pires et al. 2013).

Sensor development, especially on LoC platforms involves significant contributions from various disciplines include biochemistry, biomedicine and engineering. In general, designing a sensor itself is not enough as the biochemical reaction required for selective detection needs to be suitably designed and integrated into the detector concept. Surface modification of the sensor element is of paramount importance and various techniques for obtaining locally confined, biochemically robust, and selective modifications have been studied on various substrates including, among others, polymers (Länge et al. 2007) and glass (Waldbaur et al. 2012). Several robust detection mechanism have been extensively studied on LoC platforms some of which include electrochemical biosensors based, e.g., on impedance measurement (Guan et al. 2004), optical sensors (Guan et al. 2004) as well as gravimetric sensing (Länge et al. 2008). Often, tailored solutions have to be engineered in order to combine suitable detectors into LoC systems in a quick and robust manner and several solutions can be found in the literature (Rapp et al. 2011). We have recently summarized the most relevant detection mechanisms to which we refer to reader for further information (Rapp et al. 2010).

5.3 Case Studies

The previous chapter introduced the most important components of typical LoC systems. The following section will highlight selected applications and highlight the components used for performing the respective assays. Naturally, many more examples can be studied but one will observe that the building blocks encountered resemble or sometimes are even identical.

5.3.1 A Microfluidic Device for Genetic Analysis

In 2005 Pal et al. presented an integrated microfluidic device which can be used for a large variety of genetic assays (Pal et al. 2005). The LoC system consists of three inlet ports for sample and reagents loading, suitable channel networks for metering and guiding the various fluids, two intersections for droplet mixing, a chamber for PCR and restriction endonuclease digestion (RD) as well as a channel for electrophoresis. The system is complemented by valves for controlling the fluid flow inside of the system. For performing an assay, sample and the reagents are first loaded into the system via the inlet ports. In the next step PCR reagents and sample are mixed and transferred to the PCR chamber which is then closed by two valves. Using integrated heaters and sensors for temperature control the fluid in the PCR chamber is then thermocycled. Afterwards the valves are opened and the product is shifted into the next channel. One part of the solution is mixed with the restriction digest reaction reagent and then placed in the RD chamber. The product of this reaction is then moved to the electrophoresis section where it is separated for analysis. Pal et al. successfully identified a strain of an influenza virus with this LoC system.

5.3.2 A Microfluidic Analyzer for Measurements in Marine Waters

A portable microfluidic system for detection of soluble reactive phosphorus (SRP) in seawater was presented by Legiret et al. in 2013 (Legiret et al. 2013). The system

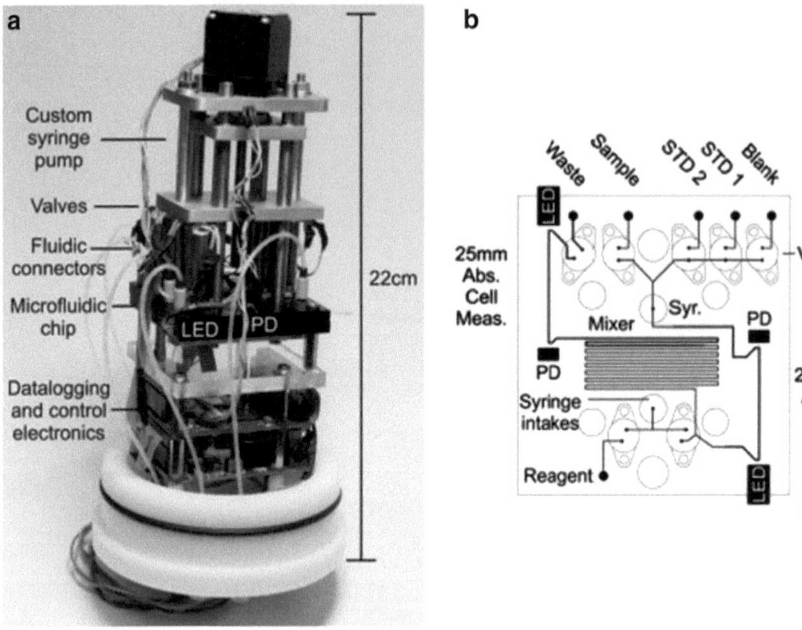

Fig. 5.9 (**a**) Photograph of the complete analyzer, (**b**) Layout of the microfluidic chip. Reprinted with permission from (Legiret et al. 2013)

combines a syringe pump, a microfluidic chip with fluid connectors and external valves and control electronics as well as a detection system (see Fig. 5.9). The microfluidic chip consists of five inlets for the sample, reagents for detection or different analysis standards, suitable channel structures, a mixing region, and photodiodes (PD) and light emitting diodes (LED) glued into the system at two detection zones (reference cell and measurement cell) for optical measurement. After injection the sample is mixed with the reagents. Because of complex formation between the orthophosphate in the seawater and the acidified vanadomolybdate reagent a color change is obtained which can be detected by absorbance measurement. A detection limit of 52 nM SRP in marine water was achieved with this system.

5.3.3 A PoC Platform for Quantifying Active Enzymes at Femtomolar Concentration Levels

In 2013 a paper-based device for quantifying active enzyme analytes on a PoC platform was presented by Lewis et al. (Lewis et al. 2013). The platform is a multilayer paper stack with several differently modified zones (see Fig. 5.10). In the center of the platform the sample is applied. The sample is then split and guided

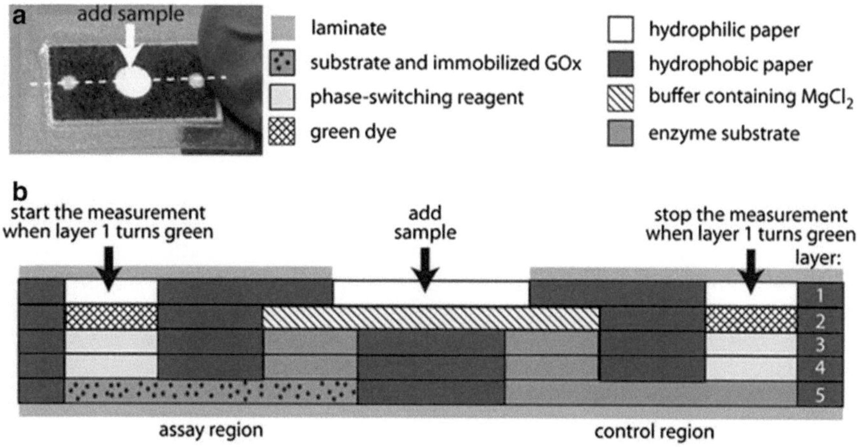

Fig. 5.10 Photograph and schematic of the device. (**a**) Picture of the paper-based microfluidic device made of stacked layers of wax-patterned paper bonded via an adhesive and then laminated. The dimensions of the device are 20 mm × 10 mm × 1.8 mm. The *white dotted line* indicates the position of the cross section shown in (**b**). (**b**) The device consists of five layers and is subdivided in an assay region on the left hand and a control region on the right hand. Adapted with permission from (Lewis et al. 2013). Copyright 2013 American Chemical Society

along two hydrophilic channels in layer 2 of the stack. The buffer serves for controlling the pH of the sample within the device. The device consists of two detection regions, both of which are connected to the sample zone by one of the hydrophilic channels, respectively. The actual measurement region is located on the left-hand side of the assay, the control region on the right-hand side. Depending on the assay to be performed, paper layers 3, 4, and 5 are partially impregnated with a suitable enzyme substrate that reacts with the analyte to be detected. This reaction is to release glucose via an enzymatic reaction. Layer 5 of the assay region contains immobilized glucose oxidase (GOx) which oxidizes the glucose and generates hydrogen peroxide. At the detection regions (both control and measurement) a phase-switching reagent is impregnated. This reagent renders the paper naturally hydrophobic and the sample can only move slowly through this region. Due to the hydrogen peroxide produced if the analyte reacted with the enzyme substrate the property of the paper can be switch back to hydrophilic if the analyte was detected. The concentration of the hydrogen oxide depends on the concentration of the target analyte and therefore controls the velocity at which the sample passes through the hydrophobic region turning hydrophilic. Passing layer 2 the sample then dissolves a dried green dye and flushes the colored solution to the top layer where it can see visually in the observation window. The control region has the same setup as the assay region except the GOx at layer 5 is missing. Therefore the channel with the phase-switching reagent remains hydrophobic due to the absence of hydrogen oxide and the sample pass slowly through this region. The time difference between the appearances of the two green spots is proportional to the concentration of the analyte. Higher time differences relate to higher quantities of the target enzyme.

For the model enzyme alkaline phosphate Lewis et al. reached a detection limit of 320 pM.

5.3.4 A PoC System for Genomic Diagnostics

A PoC system for pathogen genotyping called microfluidic biomolecular amplification reader (μBAR) was presented by Myers et al. in 2013 (Myers et al. 2013). This system (see Fig. 5.11) combines a disposable microfluidic cartridge with an

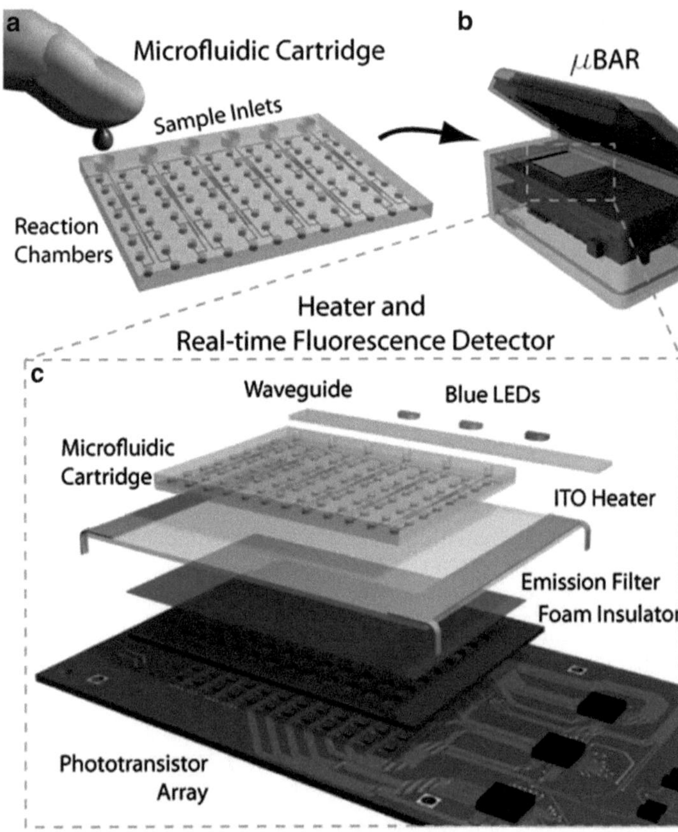

Fig. 5.11 Design and operation of the μBAR system. (**a**) The disposable microfluidic device is loaded with the sample fluid via degas-driven flow without external power or active pumping. Reagents required, e.g., primers or enzymes, have to be provided preloaded on-chip or have to be mixed with the sample before loading. (**b**) View of the complete μBAR system. (**c**) Detailed view of the different levels of the μBAR system. With help of waveguides the light of blue LEDs is coupled to the PDMS chip. An ITO heater provides constant heat distribution in the microfluidic chip. The foam insulator serves as optical isolation of each phototransistor on the circuit board and provides thermal isolation to the system. Reprinted with permission from (Myers et al. 2013)

integrated heater for temperature control in the reaction chambers of the microfluidic chip and a system to read out the fluorescence signal from each chamber by LEDs and phototransistors. The microfluidic device is made from PDMS via soft lithography and bonded to a thin polymer sheet using plasma-based bonding. The cartridge consists of 96 reaction chambers and 6 individual inlets for a multiplexed analysis. Sample loading is done by a degas-driven flow. Here the microfluidic device is first exposed to a vacuum whereupon the pores of the PDMS contract. At atmospheric pressure the chip will move the samples out of the inlets into the microfluidic structure. The assay is based on loop-mediated isothermal amplification (LAMP). For this, the microfluidic chip is placed directly above an indium tin oxide (ITO) heater. A calcein dye which is blocked by manganese ions is used as detecting probe and can be preloaded in the cartridge or can be mixed with the sample before loading. During nucleic acid amplification pyrophosphate groups are produced and drop out in form of manganese pyrophosphate which removes the quencher from the calcein. Blue indium gallium nitride (InGaN) LEDs with an emission peak at 472 nm are used to light the cartridge on the side. The calcein is excited at 480 nm wavelength and emits light at 515 nm. This fluorescence is detected by an array of phototransistors which is located below the reaction chambers. The detection limit of the μBAR is about 600 pM. Typically concentrations in nucleic acid amplification reactions are about 1–10 μM. As a first application Myers et al. demonstrate the detection of the HIV integrase gene using their platform.

5.4 Summary and Conclusion

As the work at hand demonstrates, LoC development is an active field both scientifically and commercially. Originating from classical MEMS micromachining, this discipline has developed into a multidisciplinary, application-centered scientific field and is beginning to produce both commercial products as well as generic and adaptive platforms for answering basic biological and medical questions in a high-throughput and easy-to-operate manner. As discussed, there is a wide choice of LoC platforms available nowadays with no single platform dominating. Depending on the application scope several potential solutions can be offered and systems designed to suit the need of the assay and/or the user. This chapter highlighted the most important platforms and the components commonly associated or used in conjunction with these platforms. Naturally, as new topics of research are being identified and solutions refined in terms of requirements on simplicity and assay robustness, development may favor one solution over the other. Given the wide application scope of current LoC systems and PoC devices, it seems unlikely that development on component and system design will stop in the near future. Especially given the rise (or rather rediscovery) of platforms such as the capillary flow platforms, components and solutions need to be adapted and potentially fused with existing solutions in order to tackle analytical problems at hand. Ample research on underlying fluidmechanical phenomena has been essential for this development and future improvements on technology and application development will highly profit from the extensive research on components and systems which has been done during the last three decades.

References

Abe K, Kotera K, Suzuki K, Citterio D (2010) Inkjet-printed paperfluidic immuno-chemical sensing device. Anal Bioanal Chem 398(2):885–893

Abe K, Suzuki K, Citterio D (2008) Inkjet-printed microfluidic multianalyte chemical sensing paper. Anal Chem 80(18):6928–6934

Adams ML, Johnston ML, Scherer A, Quake SR (2005) Polydimethylsiloxane based microfluidic diode. J Micromech Microeng 15(8):1517

Ahmed D, Mao X, Juluri BK, Huang TJ (2009) A fast microfluidic mixer based on acoustically driven sidewall-trapped microbubbles. Microfluid Nanofluid 7(5):727–731

Ahn CH, Allen MG (1995) Fluid micropumps based on rotary magnetic actuators. In: IEEE micro electro mechanical systems workshop (MEMS'95), Amsterdam, Netherlands. IEEE, Washington, DC, pp 408–412

Ahn CH, Jin-Woo C, Beaucage G, Nevin JH, Jeong-Bong L, Puntambekar A, Lee JY (2004) Disposable smart lab on a chip for point-of-care clinical diagnostics. Proc IEEE 92(1):154–173

Ahn S-H, Kim Y-K (1997) Fabrication and experiment of planar micro ion drag pump. In: International conference on solid state sensors and actuators, 1997 TRANSDUCERS'97 Chicago, 1997. IEEE, Washington, DC, pp 373–376

Anderson RC, Bogdan GJ, Bamiv Z, Dawes TD, Winkler J, Roy K (1997) Microfluidic bischemical analysis system. In: Transducers 1997 - international conference on solid state sensors and actuators, 16–19 Jun 1997. IEEE, Washington, DC, pp 477–480. doi:10.1109/SENSOR.1997.613690

Andersson H, van der Wijngaart W, Enoksson P, Stemme G (2000) Micromachined flow-through filter-chamber for chemical reactions on beads. Sensors Actuators B Chem 67(1–2):203–208

Andersson H, van der Wijngaart W, Griss P, Niklaus F, Stemme G (2001a) Hydrophobic valves of plasma deposited octafluorocyclobutane in DRIE channels. Sens Actuators B 75(1–2):136–141

Andersson H, van der Wijngaart W, Nilsson P, Enoksson P, Stemme G (2001b) A valve-less diffuser micropump for microfluidic analytical systems. Sens Actuators B 72(3):259–265

Andersson H, van der Wijngaart W, Stemme G (2001c) Micromachined filter-chamber array with passive valves for biochemical assays on beads. Electrophoresis 22(2):249–257

Atten P, Seyed-Yagoobi J (2003) Electrohydrodynamically induced dielectric liquid flow through pure conduction in point/plane geometry. IEEE Trans Dielectr Electr Insul 10(1):27–36

Bae B, Kee H, Kim S, Lee Y, Sim T, Kim Y, Park K (2003) In vitro experiment of the pressure regulating valve for a glaucoma implant. J Micromech Microeng 13(5):613

Baldi A, Gu YD, Loftness PE, Siegel RA, Ziaie B (2003) A hydrogel-actuated environmentally sensitive microvalve for active flow control. J Microelectromech Syst 12(5):613–621

Barbaro M, Bonfiglio A, Raffo L, Alessandrini A, Facci P, BarakBarak I (2006) A CMOS, fully integrated sensor for electronic detection of DNA hybridization. IEEE Electron Device Letters 27(7):595–597

Bart SF, Tavrow LS, Mehregany M, Lang JH (1990) Microfabricated electrohydrodynamic pumps. Sensors Actuators A Phys 21(1):193–197

Barth PW (1995) Silicon microvalves for gas flow control. In: The 8th international conference on solid-state sensors and actuators, Eurosensors IX, Transducers'95, 25–29 Jun 1995. IEEE, Washington, DC, pp 276–279. doi:10.1109/SENSOR.1995.721799

Bau HH, Zhong J, Yi M (2001) A minute magneto hydro dynamic (MHD) mixer. Sens Actuators B 79(2):207–215

Beebe DJ, Moore JS, Bauer JM, Yu Q, Liu RH, Devadoss C, Jo B-H (2000) Functional hydrogel structures for autonomous flow control inside microfluidic channels. Nature 404(6778):588–590

Belgrader P, Okuzumi M, Pourahmadi F, Borkholder DA, Northrup MA (2000) A microfluidic cartridge to prepare spores for PCR analysis. Biosens Bioelectron 14(10):849–852

Benard WL, Kahn H, Heuer AH, Huff MA (1998) Thin-film shape-memory alloy actuated micropumps. J Microelectromech Syst 7(2):245–251

Berger M, Welle A, Gottwald E, Rapp M, Länge K (2010) Biosensors coated with sulfated polysaccharides for the detection of hepatocyte growth factor/scatter factor in cell culture medium. Biosens Bioelectron 26(4):1706–1709

Berglund RN, Liu BYH (1973) Generation of monodisperse aerosol standards. Environ Sci Technol 7(2):147–153

Berthier E, Beebe DJ (2007) Flow rate analysis of a surface tension driven passive micropump. Lab Chip 7(11):1475–1478

Bessoth F, deMello AJ, Manz A (1999) Microstructure for efficient continuous flow mixing. Anal Commun 36(6):213–215

Biddiss E, Erickson D, Li D (2004) Heterogeneous surface charge enhanced micromixing for electrokinetic flows. Anal Chem 76(11):3208–3213

Bien DCS, Mitchell SJN, Gamble HS (2003) Fabrication and characterization of a micromachined passive valve. J Micromech Microeng 13(5):557

Bocong Y, Boxiong W, Werner Karl S (2010) A thermopneumatically actuated bistable microvalve. J Micromech Microeng 20(9):095024

Böhm S, Olthuis W, Bergveld P (1999a) An integrated micromachined electrochemical pump and dosing system. Biomed Microdevices 1(2):121–130

Böhm S, Olthuis W, Bergveld P (1999b) A plastic micropump constructed with conventional techniques and materials. Sensors Actuators A Phys 77(3):223–228

Bökenkamp D, Desai A, Yang X, Tai Y-C, Marzluff EM, Mayo SL (1998) Microfabricated silicon mixers for submillisecond quench-flow analysis. Anal Chem 70(2):232–236

Branebjerg J, Gravesen P, Krog JP, Nielsen CR (1996) Fast mixing by lamination. In: The ninth annual international workshop on micro electro mechanical systems, 1996, MEMS'96, Proceedings. An investigation of micro structures, sensors, actuators, machines and systems. IEEE, Washington, DC, pp 441–446

Branebjerg J, Jensen OS, Laursen NG, Leistiko O, Soeberg H (1991) A micromachined flow sensor for measuring small liquid flows. In: Transducers 1991 - international conference on solid-state sensors and actuators, 24–27 June 1991. IEEE, Washington, DC, pp 41–44. doi:10.1109/SENSOR.1991.148793

Brody JP, Yager P (1997) Diffusion-based extraction in a microfabricated device. Sensors Actuators A Phys 58(1):13–18

Burns MA, Mastrangelo CH, Sammarco TS, Man FP, Webster JR, Johnsons B, Foerster B, Jones D, Fields Y, Kaiser AR (1996) Microfabricated structures for integrated DNA analysis. Proc Natl Acad Sci 93(11):5556–5561

Burtis CA, Mailen JC, Johnson WF, Scott CD, Tiffany TO, Anderson NG (1972) Development of a miniature fast analyzer. Clin Chem 18(8):753–761

Carlen ET, Mastrangelo CH (2002) Surface micromachined paraffin-actuated microvalve. J Microelectromech Syst 11(5):408–420

Carrozza MC, Croce N, Magnani B, Dario P (1995) A piezoelectric-driven stereolithography-fabricated micropump. J Micromech Microeng 5(2):177

Clausell-Tormos J, Lieber D, Baret J-C, El-Harrak A, Miller OJ, Frenz L, Blouwolff J, Humphry KJ, Köster S, Duan H (2008) Droplet-based microfluidic platforms for the encapsulation and screening of mammalian cells and multicellular organisms. Chem Biol 15(5):427–437

Colgate E, Matsumoto H (1990) An investigation of electrowetting-based microactuation. J Vac Sci Technol A 8(4):3625–3633

Crevillén AG, Hervás M, López MA, González MC, Escarpa A (2007) Real sample analysis on microfluidic devices. Talanta 74(3):342–357

Curcio M, Roeraade J (2002) Continuous segmented-flow polymerase chain reaction for high-throughput miniaturized DNA amplification. Anal Chem 75(1):1–7

Chakraborty I, Tang WC, Bame DP, Tang TK (2000) MEMS micro-valve for space applications. Sensors Actuators A Phys 83(1–3):188–193

Chen C-H, Santiago JG (2002) A planar electroosmotic micropump. J Microelectromech Syst 11 (6):672–683

Chen IJ, Eckstein EC, Lindner E (2009) Computation of transient flow rates in passive pumping micro-fluidic systems. Lab Chip 9(1):107–114

Chen JZ, Darhuber AA, Troian SM, Wagner S (2004) Capacitive sensing of droplets for microfluidic devices based on thermocapillary actuation. Lab Chip 4(5):473–480

Chen ZY, Wang J, Qian SZ, Bau HH (2005) Thermally-actuated, phase change flow control for microfluidic systems. Lab Chip 5(11):1277–1285

Cheng C-M, Martinez AW, Gong J, Mace CR, Phillips ST, Carrilho E, Mirica KA, Whitesides GM (2010) Paper-based ELISA. Angew Chem Int Ed 49(28):4771–4774

Cheung K, Gawad S, Renaud P (2005) Impedance spectroscopy flow cytometry: on-chip label-free cell differentiation. Cytometry A 65(2):124–132

Chin CD, Linder V, Sia SK (2012) Commercialization of microfluidic point-of-care diagnostic devices. Lab Chip 12(12):2118–2134

Cho HJ, Oh KW, Ahn CH, Boolchand P, Tae-Chul N (2001) Stress analysis of silicon membranes with electroplated permalloy films using Raman scattering. IEEE Trans Magn 37(4):2749–2751

Cho ST, Wise KD (1993) A high-performance microflowmeter with built-in self test. Sensors Actuators A Phys 36(1):47–56

Choi J-W, Oh K, Han A, Wijayawardhana CA, Lannes C, Bhansali S, Schlueter K, Heineman W, Halsall HB, Nevin J, Helmicki A, Henderson HT, Ahn C (2001) Development and characterization of microfluidic devices and systems for magnetic bead-based biochemical detection. Biomed Microdevices 3(3):191–200

Choi K, Ng AH, Fobel R, Wheeler AR (2012) Digital microfluidics. Annu Rev Anal Chem 5:413–440

Chou H-P, Unger M, Quake S (2001) A microfabricated rotary pump. Biomed Microdevices 3(4):323–330

Chow AW (2002) Lab-on-a-chip: opportunities for chemical engineering. AIChE J 48(8):1590–1595

Chung S, Kim J, Wang K, Han D-C, Chang J-K (2003) Development of MEMS-based cerebrospinal fluid shunt system. Biomed Microdevices 5(4):311–321

Darabi J, Rada M, Ohadi M, Lawler J (2002) Design, fabrication, and testing of an electrohydrodynamic ion-drag micropump. J Microelectromech Syst 11(6):684–690

Dario P, Croce N, Carrozza MC, Varallo G (1996) A fluid handling system for a chemical microanalyzer. J Micromech Microeng 6(1):95

Dasgupta PK, Liu S (1994) Auxiliary electroosmotic pumping in capillary electrophoresis. Anal Chem 66(19):3060–3065

Davidsson R, Genin F, Bengtsson M, Laurell T, Emnéus J (2004) Microfluidic biosensing systems Part I. Development and optimisation of enzymatic chemiluminescent μ-biosensors based on silicon microchips. Lab Chip 4(5):481–487

de Jong J, Lammertink RGH, Wessling M (2006) Membranes and microfluidics: a review. Lab Chip 6(9):1125–1139

Deféver T, Druet M, Rochelet-Dequaire M, Joannes M, Grossiord C, Limoges B, Marchal D (2009) Real-time electrochemical monitoring of the polymerase chain reaction by mediated redox catalysis. J Am Chem Soc 131(32):11433–11441

Delapierre G (1989) Micro-machining: a survey of the most commonly used processes. Sensors Actuators 17(1–2):123–138

Dertinger SK, Chiu DT, Jeon NL, Whitesides GM (2001) Generation of gradients having complex shapes using microfluidic networks. Anal Chem 73(6):1240–1246

Deshmukh AA, Liepmann D, Pisano AP (2000) Continuous micromixer with pulsatile micropumps. In: Technical digest of the IEEE solid state sensor and actuator workshop (Hilton Head Island, SC). IEEE, Washington, DC

Dongeun H, Wei G, Yoko K, James BG, Shuichi T (2005) Microfluidics for flow cytometric analysis of cells and particles. Physiol Meas 26(3):R73

Döpper J, Clemens M, Ehrfeld W, Jung S, Kaemper K, Lehr H (1997) Micro gear pumps for dosing of viscous fluids. J Micromech Microeng 7(3):230

Du L, Zhe J (2011) A high throughput inductive pulse sensor for online oil debris monitoring. Tribol Int 44(2):175–179

Duffy DC, Gillis HL, Lin J, Sheppard NF, Kellogg GJ (1999) Microfabricated centrifugal microfluidic systems: characterization and multiple enzymatic assays. Anal Chem 71 (20):4669–4678

Duffy DC, McDonald JC, Schueller OJA, Whitesides GM (1998) Rapid prototyping of microfluidic systems in poly(dimethylsiloxane). Anal Chem 70(23):4974–4984

Easley CJ, Karlinsey JM, Bienvenue JM, Legendre LA, Roper MG, Feldman SH, Hughes MA, Hewlett EL, Merkel TJ, Ferrance JP, Landers JP (2006) A fully integrated microfluidic genetic analysis system with sample-in–answer-out capability. Proc Natl Acad Sci 103(51):19272–19277

El Moctar AO, Aubry N, Batton J (2003) Electro-hydrodynamic micro-fluidic mixer. Lab Chip 3 (4):273–280

Erickson KA, Wilding P (1993) Evaluation of a novel point-of-care system, the i-STAT portable clinical analyzer. Clin Chem 39(2):283–287

Esashi M, Shoji S, Nakano A (1989) Normally closed microvalve and mircopump fabricated on a silicon wafer. Sensors Actuators 20(1–2):163–169

Fahrenberg J, Bier W, Maas D, Menz W, Ruprecht R, Schomburg WK (1995) A microvalve system fabricated by thermoplastic molding. J Micromech Microeng 5(2):169

Fair RB (2007) Digital microfluidics: is a true lab-on-a-chip possible? Microfluid Nanofluid 3 (3):245–281

Feigl F (1935) Qualitative Analyse mit hilfe von Tuepfelreaktionen : theoretische Grundlagen, praktische Ausfuehrung und Anwendung. Akademische Verlagsgesellschaft, Leipzig

Feng G-H, Chou Y-C (2011) Fabrication and characterization of thermally driven fast turn-on microvalve with adjustable backpressure design. Microelectron Eng 88(2):187–194

Fletcher PDI, Haswell SJ, Pombo-Villar E, Warrington BH, Watts P, Wong SYF, Zhang X (2002) Micro reactors: principles and applications in organic synthesis. Tetrahedron 58(24):4735–4757

Folta JA, Raley NF, Hee EW (1992) Design, fabrication and testing of a miniature peristaltic membrane pump. In: Solid-state sensor and actuator workshop, 1992 5th technical digest, 22–25 June 1992. IEEE, Washington, DC, pp 186–189. doi:10.1109/solsen.1992.228296

Fréchette LG, Jacobson SA, Breuer KS, Ehrich FF, Ghodssi R, Khanna R, Wong CW, Zhang X, Schmidt MA, Epstein AH (2000) Demonstration of a microfabricated high-speed turbine supported on gas bearings. DTIC document

Fredrickson CK, Fan ZH (2004) Macro-to-micro interfaces for microfluidic devices. Lab Chip 4 (6):526–533

Fu C, Rummler Z, Schomburg W (2003) Magnetically driven micro ball valves fabricated by multilayer adhesive film bonding. J Micromech Microeng 13(4):S96

Fu LM, Yang RJ, Lin CH, Chien YS (2005) A novel microfluidic mixer utilizing electrokinetic driving forces under low switching frequency. Electrophoresis 26(9):1814–1824

Fuhr G (1997) From micro field cages for living cells to Brownian pumps for submicron particles. In: Proceedings of the 1997 international symposium on micromechatronics and human science, 1997. IEEE, Washington, DC, pp 1–4

Fuhr G, Hagedorn R, Muller T, Benecke W, Wagner B (1992) Microfabricated electrohydrodynamic (EHD) pumps for liquids of higher conductivity. J Microelectromech Syst 1 (3):141–146

Fuhr G, Schnelle T, Wagner B (1994) Travelling wave-driven microfabricated electrohydrodynamic pumps for liquids. J Micromech Microeng 4(4):217

Fujii T, Sando Y, Higashino K, Fujii Y (2003) A plug and play microfluidic device. Lab Chip 3 (3):193–197

Funfak A, Brösing A, Brand M, Köhler JM (2007) Micro fluid segment technique for screening and development studies on Danio rerio embryos. Lab Chip 7(9):1132–1138

Gass V, van der Schoot BH, Jeanneret S, de Rooij NF (1994) Integrated flow-regulated silicon micropump. Sensors Actuators A Phys 43(1–3):335–338

Geng X, Yuan H, Oguz HN, Prosperetti A (2001) Bubble-based micropump for electrically conducting liquids. J Micromech Microeng 11(3):270

Gerlach T (1998) Microdiffusers as dynamic passive valves for micropump applications. Sensors Actuators A Phys 69(2):181–191

Gerlach T, Wurmus H (1995) Working principle and performance of the dynamic micropump. Sensors Actuators A Phys 50(1–2):135–140

Ghindilis AL, Smith MW, Schwarzkopf KR, Roth KM, Peyvan K, Munro SB, Lodes MJ, Stöver AG, Bernards K, Dill K, McShea A (2007) CombiMatrix oligonucleotide arrays: genotyping and gene expression assays employing electrochemical detection. Biosens Bioelectron 22(9–10):1853–1860

Glasgow I, Aubry N (2003) Enhancement of microfluidic mixing using time pulsing. Lab Chip 3 (2):114–120

Go JS, Shoji S (2004) A disposable, dead volume-free and leak-free in-plane PDMS microvalve. Sensors Actuators A Phys 114(2–3):438–444

Gobby D, Angeli P, Gavriilidis A (2001) Mixing characteristics of T-type microfluidic mixers. J Micromech Microeng 11(2):126

Goetz H, Kuschel M, Wulff T, Sauber C, Miller C, Fisher S, Woodward C (2004) Comparison of selected analytical techniques for protein sizing, quantitation and molecular weight determination. J Biochem Biophys Methods 60(3):281–293

Goll C, Bacher W, Büstgens B, Maas D, Ruprecht R, Schomburg WK (1997) An electrostatically actuated polymer microvalve equipped with a movable membrane electrode. J Micromech Microeng 7(3):224

Gorkin R, Park J, Siegrist J, Amasia M, Lee BS, Park J-M, Kim J, Kim H, Madou M, Cho Y-K (2010) Centrifugal microfluidics for biomedical applications. Lab Chip 10(14):1758–1773

Gravesen P, Branebjerg J, Jensen OS (1993) Microfluidics-a review. J Micromech Microeng 3 (4):168

Greenacre CB, Flatland B, Souza MJ, Fry MM (2008) Comparison of avian biochemical test results with abaxis VetScan and Hitachi 911 analyzers. J Avian Med Surg 22(4):291–299

Grover WH, Skelley AM, Liu CN, Lagally ET, Mathies RA (2003) Monolithic membrane valves and diaphragm pumps for practical large-scale integration into glass microfluidic devices. Sens Actuators B 89(3):315–323

Gruhl FJ, Länge K (2014) Surface acoustic wave (SAW) biosensor for rapid and label-free detection of penicillin G in milk. Food Anal Methods 7(2):430–437

Gu W, Zhu X, Futai N, Cho BS, Takayama S (2004) Computerized microfluidic cell culture using elastomeric channels and Braille displays. Proc Natl Acad Sci U S A 101(45):15861–15866

Guan J-G, Miao Y-Q, Zhang Q-J (2004) Impedimetric biosensors. J Biosci Bioeng 97(4):219–226

Gui L, Liu J (2004) Ice valve for a mini/micro flow channel. J Micromech Microeng 14(2):242–246

Gui L, Yu BY, Ren CL, Huissoon JP (2011) Microfluidic phase change valve with a two-level cooling/heating system. Microfluid Nanofluid 10(2):435–445

Guttenberg Z, Müller H, Habermüller H, Geisbauer A, Pipper J, Felbel J, Kielpinski M, Scriba J, Wixforth A (2005) Planar chip device for PCR and hybridization with surface acoustic wave pump. Lab Chip 5(3):308–317

Haeberle S, Zengerle R (2007) Microfluidic platforms for lab-on-a-chip applications. Lab Chip 7 (9):1094–1110

Handique K, Burke D, Mastrangelo C, Burns M (2001) On-chip thermopneumatic pressure for discrete drop pumping. Anal Chem 73(8):1831–1838

Hannig C, Dirschka M, Länge K, Neumaier S, Rapp BE (2010) Synthesis and application of photo curable perfluoropolyethers as new material for microfluidics. Procedia Engineering 5:866–869

Hao R, Wang D, Zhang X, Zuo G, Wei H, Yang R, Zhang Z, Cheng Z, Guo Y, Cui Z (2009) Rapid detection of *Bacillus anthracis* using monoclonal antibody functionalized QCM sensor. Biosens Bioelectron 24(5):1330–1335

Hartshorne H, Backhouse CJ, Lee WE (2004) Ferrofluid-based microchip pump and valve. Sens Actuators B 99(2–3):592–600

Hasegawa T, Nakashima K, Omatsu F, Ikuta K (2008) Multi-directional micro-switching valve chip with rotary mechanism. Sensors Actuators A Phys 143(2):390–398

Hatch A, Kamholz AE, Holman G, Yager P, Bohringer KF (2001) A ferrofluidic magnetic micropump. J Microelectromech Syst 10(2):215–221

He B, Burke BJ, Zhang X, Zhang R, Regnier FE (2001) A picoliter-volume mixer for microfluidic analytical systems. Anal Chem 73(9):1942–1947

Heckele M, Schomburg WK (2004) Review on micro molding of thermoplastic polymers. J Micromech Microeng 14(3):R1–R14

Hessel V, Hardt S, Löwe H, Schönfeld F (2003) Laminar mixing in different interdigital micromixers: I. Experimental characterization. AIChE J 49(3):566–577

Hinsmann P, Frank J, Svasek P, Harasek M, Lendl B (2001) Design, simulation and application of a new micromixing device for time resolved infrared spectroscopy of chemical reactions in solution. Lab Chip 1(1):16–21

Hong C-C, Chang P-H, Lin C-C, Hong C-L (2010a) A disposable microfluidic biochip with on-chip molecularly imprinted biosensors for optical detection of anesthetic propofol. Biosens Bioelectron 25(9):2058–2064

Hong C-C, Choi J-W, Ahn C (2001) A novel in-plane passive micromixer using Coanda effect. In: Ramsey JM, Berg A (eds) Micro total analysis systems. Springer, Dordrecht, The Netherlands, pp 31–33. doi:10.1007/978-94-010-1015-3_11

Hong C-C, Choi J-W, Ahn CH (2004) A novel in-plane passive microfluidic mixer with modified Tesla structures. Lab Chip 4(2):109–113

Hong J, Choi JS, Han G, Kang JK, Kim C-M, Kim TS, Yoon DS (2006) A Mach-Zehnder interferometer based on silicon oxides for biosensor applications. Anal Chim Acta 573–574:97–103

Hong T-F, Ju W-J, Wu M-C, Tai C-H, Tsai C-H, Fu L-M (2010b) Rapid prototyping of PMMA microfluidic chips utilizing a CO2 laser. Microfluid Nanofluid 9(6):1125–1133

Horade M, Mizuta Y, Kaji N, Higashiyama T, Arata H (2012) Plant-on-a-chip microfluidic-system for quantitative analysis of pollen tube guidance by signaling molecule: towards cell-to-cell communication study. In: Proc microTAS, 2012, pp 1027–1029

Hua SZ, Sachs F, Yang DX, Chopra HD (2002) Microfluidic actuation using electrochemically generated bubbles. Anal Chem 74(24):6392–6396

Huang LR, Cox EC, Austin RH, Sturm JC (2004) Continuous particle separation through deterministic lateral displacement. Science 304(5673):987–990

Huang M-Z, Yang R-J, Tai C-H, Tsai C-H, Fu L-M (2006) Application of electrokinetic instability flow for enhanced micromixing in cross-shaped microchannel. Biomed Microdevices 8 (4):309–315

Huang S-B, Wu M-H, Cui Z, Cui Z, Lee G-B (2008) A membrane-based serpentine-shape pneumatic micropump with pumping performance modulated by fluidic resistance. J Micromech Microeng 18(4):045008

Huang X, Gordon MJ, Zare RN (1988) Current-monitoring method for measuring the electroosmotic flow rate in capillary zone electrophoresis. Anal Chem 60(17):1837–1838

Huh YS, Choi JH, Huh KA, Park TJ, Hong YK, Kim do H, Hong WH, Lee SY (2007) Microfluidic cell disruption system employing a magnetically actuated diaphragm. Electrophoresis 28 (24):4748–4757

Jang J, Lee SS (2000) Theoretical and experimental study of MHD (magnetohydrodynamic) micropump. Sensors Actuators A Phys 80(1):84–89

Jen C-P, Lin Y-C (2002) Design and simulation of bi-directional microfluid driving systems. J Micromech Microeng 12(2):115

Jensen K (1998) Chemical kinetics: smaller, faster chemistry. Nature 393(6687):735–737

Jensen KF (2006) Silicon-based microchemical systems: characteristics and applications. MRS Bull 31(02):101–107

Jeon N, Chiu D, Wargo C, Wu H, Choi I, Anderson J, Whitesides G (2002) Microfluidics section: design and fabrication of integrated passive valves and pumps for flexible polymer 3-dimensional microfluidic systems. Biomed Microdevices 4(2):117–121

Jeong OC, Konishi S (2008) Fabrication of a peristaltic micro pump with novel cascaded actuators. J Micromech Microeng 18(2):025022

Jeong OC, Yang SS (2000) Fabrication and test of a thermopneumatic micropump with a corrugated p+ diaphragm. Sensors Actuators A Phys 83(1–3):249–255

Jeong S-i, Seyed-Yagoobi J (2002) Experimental study of electrohydrodynamic pumping through conduction phenomenon. J Electrost 56(2):123–133

Jerman H (1994) Electrically activated normally closed diaphragm valves. J Micromech Microeng 4(4):210

Jiang F, Drese K, Hardt S, Küpper M, Schönfeld F (2004) Helical flows and chaotic mixing in curved micro channels. AIChE J 50(9):2297–2305

Johnson RD, Badr IHA, Barrett G, Lai S, Lu Y, Madou MJ, Bachas LG (2001) Development of a fully integrated analysis system for ions based on ion-selective optodes and centrifugal microfluidics. Anal Chem 73(16):3940–3946

Johnston I, Tracey M, Davis J, Tan C (2005) Microfluidic solid phase suspension transport with an elastomer-based, single piezo-actuator, micro throttle pump. Lab Chip 5(3):318–325

Joo B-S, Huh J-S, Lee D-D (2007) Fabrication of polymer SAW sensor array to classify chemical warfare agents. Sens Actuators B 121(1):47–53

Jorgenson JW, Lukacs KD (1981) Free-zone electrophoresis in glass capillaries. Clin Chem 27 (9):1551–1553

Ju W-J, Fu L-M, Yang R-J, Lee C-L (2012) Distillation and detection of SO2 using a microfluidic chip. Lab Chip 12(3):622–626

Judy JW, Tamagawa T, Polla DL (1991) Surface-machined micromechanical membrane pump. In: Micro electro mechanical systems, 1991, MEMS'91, proceedings An investigation of micro structures, sensors, actuators, machines and robots, 30 Jan–2 Feb 1991. IEEE, Washington, DC, pp 182–186. doi:10.1109/memsys.1991.114792

Jun TK (1998) Valveless pumping using traversing vapor bubbles in microchannels. J Appl Phys 83(11):5658–5664

Juncker D, Schmid H, Drechsler U, Wolf H, Wolf M, Michel B, de Rooij N, Delamarche E (2002) Autonomous microfluidic capillary system. Anal Chem 74(24):6139–6144

Kahn H, Huff MA, Heuer AH (1998) The TiNi shape-memory alloy and its applications for MEMS. J Micromech Microeng 8(3):213

Kang TG, Kwon TH (2004) Colored particle tracking method for mixing analysis of chaotic micromixers. J Micromech Microeng 14(7):891

Kataoka DE, Troian SM (1999) Patterning liquid flow on the microscopic scale. Nature 402 (6763):794–797

Kawakatsu T, Kikuchi Y, Nakajima M (1997) Regular-sized cell creation in microchannel emulsification by visual microprocessing method. J Amer Oil Chem Soc 74(3):317–321

Kawakatsu T, Trägårdh G, Kikuchi Y, Nakajima M, Komori H, Yonemoto T (2000) Effect of microchannel structure on droplet size during crossflow microchannel emulsification. J Surfact Deterg 3(3):295–302

Kazuo H, Ryutaro M (2000) A pneumatically-actuated three-way microvalve fabricated with polydimethylsiloxane using the membrane transfer technique. J Micromech Microeng 10(3):415

Khan MF, Schmid S, Larsen PE, Davis ZJ, Yan W, Stenby EH, Boisen A (2013) Online measurement of mass density and viscosity of pL fluid samples with suspended microchannel resonator. Sens Actuators B 185:456–461

Khoo M, Liu C (2001) Micro magnetic silicone elastomer membrane actuator. Sensors Actuators A Phys 89(3):259–266

Kim DS, Lee SW, Kwon TH, Lee SS (2004) A barrier embedded chaotic micromixer. J Micromech Microeng 14(6):798

Kim J, Baek J, Lee K, Park Y, Sun K, Lee T, Lee S (2006) Photopolymerized check valve and its integration into a pneumatic pumping system for biocompatible sample delivery. Lab Chip 6 (8):1091–1094

Kim J, Byun D, Mauk MG, Bau HH (2009) A disposable, self-contained PCR chip. Lab Chip 9 (4):606–612

Kim P, Kwon KW, Park MC, Lee SH, Kim SM, Suh KY (2008) Soft lithography for microfluidics: a review. Biochip Journal 2:1–11

Kirby BJ, Shepodd TJ, Hasselbrink EF Jr (2002) Voltage-addressable on/off microvalves for high-pressure microchip separations. J Chromatogr A 979(1–2):147–154

Klintberg L, Karlsson M, Stenmark L, Schweitz J-Å, Thornell G (2002) A large stroke, high force paraffin phase transition actuator. Sensors Actuators A Phys 96(2–3):189–195

Knight JB, Vishwanath A, Brody JP, Austin RH (1998) Hydrodynamic focusing on a silicon chip: mixing nanoliters in microseconds. Phys Rev Lett 80(17):3863

Koch M, Chatelain D, Evans AGR, Brunnschweiler A (1998) Two simple micromixers based on silicon. J Micromech Microeng 8(2):123

Koch M, Evans AGR, Brunnschweiler A (1997) Characterization of micromachined cantilever valves. J Micromech Microeng 7(3):221

Koch M, Witt H, Evans A, Brunnschweiler A (1999) Improved characterization technique for micromixers. J Micromech Microeng 9(2):156

Kohl M, Dittmann D, Quandt E, Winzek B (2000) Thin film shape memory microvalves with adjustable operation temperature. Sensors Actuators A Phys 83(1–3):214–219

Kohl M, Dittmann D, Quandt E, Winzek B, Miyazaki S, Allen DM (1999a) Shape memory microvalves based on thin films or rolled sheets. Mater Sci Eng A 273–275:784–788

Kohl M, Skrobanek KD, Miyazaki S (1999b) Development of stress-optimised shape memory microvalves. Sensors Actuators A Phys 72(3):243–250

Kohlrausch F (1897) Ueber Concentrations-Verschiebungen durch Elektrolyse im Innern von Lösungen und Lösungsgemischen. Ann Phys Chem 62:209–239

Kopf-Sill AR (2002) PROFILESuccesses and challenges of lab-on-a-chip. Lab Chip 2(3):42N–47N

Kortmann H, Blank LM, Schmid A (2011) Single cell analytics: an overview. In: High resolution microbial single cell analytics. Springer, Dordrecht, The Netherlands, pp 99–122

Kurosawa M, Watanabe T, Higuchi T (1995) Surface acoustic wave atomizer with pumping effect. In: Micro electro mechanical systems. IEEE, Washington, DC

Kwang-Seok Y, Il-Joo C, Bu J-U, Chang-Jin K, Euisik Y (2002) A surface-tension driven micropump for low-voltage and low-power operations. J Microelectromech Syst 11(5):454–461

Lagally E, Medintz I, Mathies R (2001) Single-molecule DNA amplification and analysis in an integrated microfluidic device. Anal Chem 73(3):565–570

Lagally ET, Simpson PC, Mathies RA (2000) Monolithic integrated microfluidic DNA amplification and capillary electrophoresis analysis system. Sens Actuators B 63(3):138–146

Länge K, Grimm S, Rapp M (2007) Chemical modification of parylene C coatings for SAW biosensors. Sens Actuators B 125(2):441–446

Länge K, Rapp BE, Rapp M (2008) Surface acoustic wave biosensors: a review. Anal Bioanal Chem 391(5):1509–1519

Lao AI, Lee TM, Hsing I, Ip NY (2000) Precise temperature control of microfluidic chamber for gas and liquid phase reactions. Sensors Actuators A Phys 84(1):11–17

Laser DJ, Goodson KE, Santiago JG, Kenny TW (2002) High-frequency actuation with silicon electroosmotic micropumps. In: Proc 2002 solid-state sensor, actuator, and microsystems workshop (Hilton Head Island, SC). IEEE, Washington, DC

Lee BS, Lee J-N, Park J-M, Lee J-G, Kim S, Cho Y-K, Ko C (2009) A fully automated immunoassay from whole blood on a disc. Lab Chip 9(11):1548–1555

Lee J, Moon H, Fowler J, Schoellhammer T, Kim C-J (2002) Electrowetting and electrowetting-on-dielectric for microscale liquid handling. Sensors Actuators A Phys 95(2–3):259–268

Lee S, Jeong O, Yang S (1998) The fabrication of a micro injector actuated by boiling and/or electrolysis. In: The eleventh annual international workshop on micro electro mechanical systems, 1998, MEMS 98. Proceedings. IEEE, Washington, DC, pp 51–56

Lefèvre F, Chalifour A, Yu L, Chodavarapu V, Juneau P, Izquierdo R (2012) Algal fluorescence sensor integrated into a microfluidic chip for water pollutant detection. Lab Chip 12(4):787–793

Legendre LA, Bienvenue JM, Roper MG, Ferrance JP, Landers JP (2006) A simple, valveless microfluidic sample preparation device for extraction and amplification of DNA from nanoliter-volume samples. Anal Chem 78(5):1444–1451

Legiret F-E, Sieben VJ, Woodward EMS, Abi Kaed Bey SK, Mowlem MC, Connelly DP, Achterberg EP (2013) A high performance microfluidic analyser for phosphate measurements in marine waters using the vanadomolybdate method. Talanta 116:382–387

Lemoff AV, Lee AP (2000) An AC magnetohydrodynamic micropump. Sens Actuators B 63 (3):178–185

Lemoff AV, Lee AP, Miles RR, McConaghy CF (1999) An AC magnetohydrodynamic micropump: towards a true integrated microfluidic system. In: 10th International conference on solid-state sensors and actuator, 1999. IEEE, Washington, DC, pp 1126–1129

Lewis GG, Robbins JS, Phillips ST (2013) Point-of-care assay platform for quantifying active enzymes to femtomolar levels using measurements of time as the readout. Anal Chem 85 (21):10432–10439

Li B, Chen Q, Lee D-G, Woolman J, Carman GP (2005) Development of large flow rate, robust, passive micro check valves for compact piezoelectrically actuated pumps. Sensors Actuators A Phys 117(2):325–330

Li H, Roberts D, Steyn J, Turner K, Carretero J, Yaglioglu O, Su Y, Saggere L, Hagood N, Spearing S (2000) A high frequency high flow rate piezoelectrically driven MEMS micropump. In: Proceedings IEEE solid state sensors and actuators workshop, Hilton Head. IEEE, Washington, DC

Li X, Tian J, Garnier G, Shen W (2010) Fabrication of paper-based microfluidic sensors by printing. Colloids Surf B Biointerfaces 76(2):564–570

Lichtenberg J, de Rooij NF, Verpoorte E (2002) Sample pretreatment on microfabricated devices. Talanta 56(2):233–266

Lien K-Y, Lee W-C, Lei H-Y, Lee G-B (2007) Integrated reverse transcription polymerase chain reaction systems for virus detection. Biosens Bioelectron 22(8):1739–1748

Lin C-F, Lee G-B, Wang C-H, Lee H-H, Liao W-Y, Chou T-C (2006) Microfluidic pH-sensing chips integrated with pneumatic fluid-control devices. Biosens Bioelectron 21(8):1468–1475

Lin C-H, Wang Y-N, Fu L-M (2012) Integrated microfluidic chip for rapid DNA digestion and time-resolved capillary electrophoresis analysis. Biomicrofluidics 6(1):012818

Lin T-Y, Hu C-H, Chou T-C (2004) Determination of albumin concentration by MIP-QCM sensor. Biosens Bioelectron 20(1):75–81

Liu B-F, Ozaki M, Hisamoto H, Luo Q, Utsumi Y, Hattori T, Terabe S (2004a) Microfluidic chip toward cellular ATP and ATP-conjugated metabolic analysis with bioluminescence detection. Anal Chem 77(2):573–578

Liu L, Chen X, Niu X, Wen W, Sheng P (2006) Electrorheological fluid-actuated microfluidic pump. Appl Phys Lett 89(8):083505-083505–083503

Liu RH, Bonanno J, Yang J, Lenigk R, Grodzinski P (2004b) Single-use, thermally actuated paraffin valves for microfluidic applications. Sens Actuators B 98(2–3):328–336

Liu RH, Stremler MA, Sharp KV, Olsen MG, Santiago JG, Adrian RJ, Aref H, Beebe DJ (2000) Passive mixing in a three-dimensional serpentine microchannel. J Microelectromech Syst 9 (2):190–197

Liu RH, Yang J, Lenigk R, Bonanno J, Grodzinski P (2004c) Self-contained, fully integrated biochip for sample preparation, polymerase chain reaction amplification, and DNA microarray detection. Anal Chem 76(7):1824–1831

Liu RH, Yu Q, Beebe DJ (2002) Fabrication and characterization of hydrogel-based microvalves. J Microelectromech Syst 11(1):45–53

London A, Epstein A, Kerrebrock J (2001) High-pressure bipropellant microrocket engine. J Propuls Power 17(4):780–787

Lu L-H, Ryu K, Liu C (2001) A novel microstirrer and arrays for microfluidic mixing. In: Ramsey JM, Berg A (eds) Micro total analysis systems 2001. Springer, Dordrecht, The Netherlands, pp 28–30. doi:10.1007/978-94-010-1015-3_10

Lu L-H, Ryu KS, Liu C (2002) A magnetic microstirrer and array for microfluidic mixing. J Microelectromech Syst 11(5):462–469

Lui C, Stelick S, Cady N, Batt C (2010) Low-power microfluidic electro-hydraulic pump (EHP). Lab Chip 10(1):74–79

Luque A, Quero JM, Hibert C, Flückiger P, Gañán-Calvo AM (2005) Integrable silicon microfluidic valve with pneumatic actuation. Sensors Actuators A Phys 118(1):144–151

Lloyd DK (1996) Capillary electrophoretic analyses of drugs in body fluids: sample pretreatment and methods for direct injection of biofluids. J Chromatogr A 735(1–2):29–42

Madou MJ, Kellogg GJ (1998) LabCD: a centrifuge-based microfluidic platform for diagnostics. In: Proc. SPIE 3259, systems and technologies for clinical diagnostics and drug discovery, pp 80–93

Marseille O, Habib N, Reul H, Rau G (1998) Implantable micropump system for augmented liver perfusion. Artif Organs 22(6):458–460

Martinez AW, Phillips ST, Butte MJ, Whitesides GM (2007) Patterned paper as a platform for inexpensive, low-volume, portable bioassays. Angew Chem Int Ed 46(8):1318–1320

Martinez AW, Phillips ST, Whitesides GM (2008) Three-dimensional microfluidic devices fabricated in layered paper and tape. Proc Natl Acad Sci 105(50):19606–19611

Mason TG, Bibette J (1997) Shear rupturing of droplets in complex fluids. Langmuir 13(17):4600–4613

McKnight TE, Culbertson CT, Jacobson SC, Ramsey JM (2001) Electroosmotically induced hydraulic pumping with integrated electrodes on microfluidic devices. Anal Chem 73 (16):4045–4049

Meckes A, Behrens J, Kayser O, Benecke W, Becker T, Müller G (1999) Microfluidic system for the integration and cyclic operation of gas sensors. Sensors Actuators A Phys 76(1–3):478–483

Mehta G, Mehta K, Sud D, Song J, Bersano-Begey T, Futai N, Heo YS, Mycek M-A, Linderman J, Takayama S (2007) Quantitative measurement and control of oxygen levels in microfluidic poly(dimethylsiloxane) bioreactors during cell culture. Biomed Microdevices 9(2):123–134

Melin J, Giménez G, Roxhed N, van der Wijngaart W, Stemme G (2004a) A fast passive and planar liquid sample micromixer. Lab Chip 4(3):214–219

Melin J, Roxhed N, Gimenez G, Griss P, van der Wijngaart W, Stemme G (2004b) A liquid-triggered liquid microvalve for on-chip flow control. Sens Actuators B 100(3):463–468

Mengeaud V, Josserand J, Girault HH (2002) Mixing processes in a zigzag microchannel: finite element simulations and optical study. Anal Chem 74(16):4279–4286

Meyvantsson I, Warrick JW, Hayes S, Skoien A, Beebe DJ (2008) Automated cell culture in high density tubeless microfluidic device arrays. Lab Chip 8(5):717–724

Mikkers FEP, Everaerts FM, Verheggen TPEM (1979) High-performance zone electrophoresis. J Chromatogr A 169:11–20

Minas G, Martins JS, Ribeiro JC, Wolffenbuttel RF, Correia JH (2004) Biological microsystem for measuring uric acid in biological fluids. Sensors Actuators A Phys 110(1–3):33–38

Mirica KA, Weis JG, Schnorr JM, Esser B, Swager TM (2012) Mechanical drawing of gas sensors on paper. Angew Chem Int Ed 51(43):10740–10745

Miyake R, Tsuzuki K, Takagi T, Imai K (1997) A highly sensitive and small flow-type chemical analysis system with integrated absorptiometric micro-flowcell. In: Tenth annual international workshop on micro electro mechanical systems, 1997, MEMS'97, Proceedings. IEEE, Washington, DC, pp 102–107

Mizoguchi H, Ando M, Mizuno T, Takagi T, Nakajima N (1992) Design and fabrication of light driven micropump. In: Micro electro mechanical systems, 1992, MEMS'92, Proceedings. An investigation of micro structures, sensors, actuators, machines and robot, 4–7 Feb 1992. IEEE, Washington, DC, pp 31–36. doi:10.1109/memsys.1992.187686

Moroney R, White R, Howe R (1991) Ultrasonically induced microtransport. In: Micro electro mechanical systems, 1991, MEMS'91, Proceedings. An investigation of micro structures, sensors, actuators, machines and robots. IEEE, Washington, DC, pp 277–282

Münchow G, Dadic D, Doffing F, Hardt S, Drese K-S (2005) Automated chip-based device for simple and fast nucleic acid amplification. Expert Rev Mol Diagn 5(4):613–620

Munson MS, Yager P (2004) Simple quantitative optical method for monitoring the extent of mixing applied to a novel microfluidic mixer. Anal Chim Acta 507(1):63–71

Myers FB, Henrikson RH, Bone J, Lee LP (2013) A handheld point-of-care genomic diagnostic system. PLoS One 8(8):e70266

Nagrath S, Sequist LV, Maheswaran S, Bell DW, Irimia D, Ulkus L, Smith MR, Kwak EL, Digumarthy S, Muzikansky A (2007) Isolation of rare circulating tumour cells in cancer patients by microchip technology. Nature 450(7173):1235–1239

Nakashima T, Shimizu M (1993) Preparation of monodispersed O/W emulsion by porous glass membrane. Kagaku Kogaku Ronbunshu 19:984

Neagu CR, Gardeniers GE, Elwenspoek M, Kelly JJ (1996) An electrochemical microactuator: principle and first results. J Microelectromech Syst 5(1):2–9

Neumann C, Voigt A, Pires L, Rapp BE (2013) Design and characterization of a platform for thermal actuation of up to 588 microfluidic valves. Microfluid Nanofluid 14(1–2):177–186

Newman JD, Turner APF (2005) Home blood glucose biosensors: a commercial perspective. Biosens Bioelectron 20(12):2435–2453

Nguyen N-T, Huang X (2001) Miniature valveless pumps based on printed circuit board technique. Sensors Actuators A Phys 88(2):104–111

Nguyen N-T, Meng AH, Black J, White RM (2000) Integrated flow sensor for in situ measurement and control of acoustic streaming in flexural plate wave micropumps. Sensors Actuators A Phys 79(2):115–121

Nguyen N-T, Wu Z (2005) Micromixers—a review. J Micromech Microeng 15(2):R1

Nguyen TT, Goo NS, Nguyen VK, Yoo Y, Park S (2008) Design, fabrication, and experimental characterization of a flap valve IPMC micropump with a flexibly supported diaphragm. Sensors Actuators A Phys 141(2):640–648

Niu X, Lee Y-K (2003) Efficient spatial-temporal chaotic mixing in microchannels. J Micromech Microeng 13(3):454

Norbert S, Thomas F, Helmut W (1996) A modular microfluid system with an integrated micromixer. J Micromech Microeng 6(1):99

Oddy M, Santiago J, Mikkelsen J (2001) Electrokinetic instability micromixing. Anal Chem 73 (24):5822–5832

Ogden S, Boden R, Hjort K (2010) A latchable valve for high-pressure microfluidics. J Microelectromech Syst 19(2):396–401

Oh KW, Ahn CH (2006) A review of microvalves. J Micromech Microeng 16(5):R13–R39

Oh KW, Park C, Namkoong K, Kim J, Ock K-S, Kim S, Kim Y-A, Cho Y-K, Ko C (2005) World-to-chip microfluidic interface with built-in valves for multichamber chip-based PCR assays. Lab Chip 5(8):845–850

Ohori T, Shoji S, Miura K, Yotsumoto A (1998) Partly disposable three-way microvalve for a medical micro total analysis system (μTAS). Sensors Actuators A Phys 64(1):57–62

Olsson A, Stemme G, Stemme E (1995) A valve-less planar fluid pump with two pump chambers. Sensors Actuators A Phys 47(1–3):549–556

Olsson A, Stemme G, Stemme E (2000) Numerical and experimental studies of flat-walled diffuser elements for valve-less micropumps. Sensors Actuators A Phys 84(1–2):165–175

Pal R, Yang M, Johnson BN, Burke DT, Burns MA (2004) Phase change microvalve for integrated devices. Anal Chem 76(13):3740–3748

Pal R, Yang M, Lin R, Johnson B, Srivastava N, Razzacki S, Chomistek K, Heldsinger D, Haque R, Ugaz V (2005) An integrated microfluidic device for influenza and other genetic analyses. Lab Chip 5(10):1024–1032

Pamme N (2007) Continuous flow separations in microfluidic devices. Lab Chip 7(12):1644–1659

Papageorgiou DT (1995) On the breakup of viscous liquid threads. Physics of Fluids (1994-present) 7(7):1529–1544

Park S-J, Kim JK, Park J, Chung S, Chung C, Chang JK (2004) Rapid three-dimensional passive rotation micromixer using the breakup process. J Micromech Microeng 14(1):6

Peige S, Zeno R, Werner Karl S (2004) Polymer micro piezo valve with a small dead volume. J Micromech Microeng 14(2):305

Peirs J, Reynaerts D, Van Brussel H (2000) Design of miniature parallel manipulators for integration in a self-propelling endoscope. Sensors Actuators A Phys 85(1–3):409–417

Petersen NJ, Mogensen KB, Kutter JP (2002) Performance of an in-plane detection cell with integrated waveguides for UV/Vis absorbance measurements on microfluidic separation devices. Electrophoresis 23(20):3528–3536

Petralia S, Verardo R, Klaric E, Cavallaro S, Alessi E, Schneider C (2013) In-Check system: a highly integrated silicon Lab-on-Chip for sample preparation, PCR amplification and microarray detection of nucleic acids directly from biological samples. Sens Actuators B 187:99–105

Pfleging W, Torge M, Bruns M, Trouillet V, Welle A, Wilson S (2009) Laser- and UV-assisted modification of polystyrene surfaces for control of protein adsorption and cell adhesion. Appl Surf Sci 255(10):5453–5457

Pickup JC, Shaw GW, Claremont DJ (1989) In vivo molecular sensing in diabetes mellitus: an implantable glucose sensor with direct electron transfer. Diabetologia 32(3):213–217

Pires L, Sachsenheimer K, Kleintschek T, Waldbaur A, Schwartz T, Rapp BE (2013) Online monitoring of biofilm growth and activity using a combined multi-channel impedimetric and amperometric sensor. Biosens Bioelectron 47:157–163

Piruska A, Nikcevic I, Lee SH, Ahn C, Heineman WR, Limbach PA, Seliskar CJ (2005) The autofluorescence of plastic materials and chips measured under laser irradiation. Lab Chip 5 (12):1348–1354

Pollack M, Shenderov A, Fair R (2002) Electrowetting-based actuation of droplets for integrated microfluidics. Lab Chip 2(2):96–101

Pu Q, Oyesanya O, Thompson B, Liu S, Alvarez JC (2006) On-chip micropatterning of plastic (Cylic Olefin Copolymer, COC) microfluidic channels for the fabrication of biomolecule microarrays using photografting methods. Langmuir 23(3):1577–1583

Puckett LG, Dikici E, Lai S, Madou M, Bachas LG, Daunert S (2004) Investigation into the applicability of the centrifugal microfluidics platform for the development of protein–ligand binding assays incorporating enhanced green fluorescent protein as a fluorescent reporter. Anal Chem 76(24):7263–7268

Ramsey R, Ramsey J (1997) Generating electrospray from microchip devices using electroosmotic pumping. Anal Chem 69(6):1174–1178

Rapp BE, Carneiro L, Laenge K, Rapp M (2009) An indirect microfluidic flow injection analysis (FIA) system allowing diffusion free pumping of liquids by using tetradecane as intermediary liquid. Lab Chip 9(2):354–356

Rapp BE, Gruhl FJ, Länge K (2010) Biosensors with label-free detection designed for diagnostic applications. Anal Bioanal Chem 398(6):2403–2412

Rapp BE, Schickling B, Prokop J, Piotter V, Rapp M, Laenge K (2011) Design and integration of a generic disposable array-compatible sensor housing into an integrated disposable indirect microfluidic flow injection analysis system. Biomed Microdevices 13(5):909–922

Rapp R, Schomburg WK, Maas D, Schulz J, Stark W (1994) LIGA micropump for gases and liquids. Sensors Actuators A Phys 40(1):57–61

Rasmussen A, Zaghloul ME (1999) The design and fabrication of microfluidic flow sensors. In: Proceedings of the 1999 I.E. international symposium on circuits and systems, 1999, ISCAS'99. IEEE, Washington, DC, pp 136–139

Recknor JB, Sakaguchi DS, Mallapragada SK (2006) Directed growth and selective differentiation of neural progenitor cells on micropatterned polymer substrates. Biomaterials 27(22):4098–4108

Rehm JE, Shepodd TJ, Hasselbrink EF (2001) Mobile flow control elements for high-pressure micro-analytical systems fabricated using in-situ polymerization. In: Ramsey JM, Berg A (eds) Micro total analysis systems. Springer, Dordrecht, The Netherlands, pp 227–229. doi:10.1007/978-94-010-1015-3_98

Rich CA, Wise KD (2003) A high-flow thermopneumatic microvalve with improved efficiency and integrated state sensing. J Microelectromech Syst 12(2):201–208

Richter A, Kuckling D, Howitz S, Gehring T, Arndt KF (2003) Electronically controllable microvalves based on smart hydrogels: magnitudes and potential applications. J Microelectromech Syst 12(5):748–753

Richter A, Plettner A, Hofmann K, Sandmaier H (1991) A micromachined electrohydrodynamic (EHD) pump. Sensors Actuators A Phys 29(2):159–168

Richter A, Sandmaier H (1990) An electrohydrodynamic micropump. In: Micro electro mechanical systems, 1990 Proceedings. An investigation of micro structures, sensors, actuators, machines and robots. IEEE, Washington, DC, pp 99–104

Rife J, Bell M, Horwitz J, Kabler M, Auyeung R, Kim W (2000) Miniature valveless ultrasonic pumps and mixers. Sensors Actuators A Phys 86(1):135–140

Roberts DC, Hanqing L, Steyn JL, Yaglioglu O, Spearing SM, Schmidt MA, Hagood NW (2003) A piezoelectric microvalve for compact high-frequency, high-differential pressure hydraulic micropumping systems. J Microelectromech Syst 12(1):81–92

Rogge T, Rummler Z, Schomburg WK (2004) Polymer micro valve with a hydraulic piezo-drive fabricated by the AMANDA process. Sensors Actuators A Phys 110(1–3):206–212

Ross D, Gaitan M, Locascio L (2001) Temperature measurement and control in microfluidic systems. In: Ramsey JM, Berg A (eds) Micro total analysis systems. Springer, Dordrecht, The Netherlands, pp 239–241. doi:10.1007/978-94-010-1015-3_102

Ryu S, Yoo I, Song S, Yoon B, Kim J-M (2009) A thermoresponsive fluorogenic conjugated polymer for a temperature sensor in microfluidic devices. J Am Chem Soc 131(11):3800–3801

Saarela V, Franssila S, Tuomikoski S, Marttila S, Ostman P, Sikanen T, Kotiaho T, Kostiainen R (2006) Re-usable multi-inlet PDMS fluidic connector. Sensors Actuators B Chem 114(1):552–557

Sabourin D, Snakenborg D, Dufva M (2009) Interconnection blocks: a method for providing reusable, rapid, multiple, aligned and planar microfluidic interconnections. J Micromech Microeng 19(3):035021

Sadler DJ, Oh KW, Ahn CH, Bhansali S, Henderson HT (1999) A new magnetically actuated microvalve for liquid and gas control applications. In: Proceedings of Transducers, 1999. pp 1812–1815

Sammarco TS, Burns MA (1999) Thermocapillary pumping of discrete drops in microfabricated analysis devices. AIChE J 45(2):350–366

Santiago JG, Wereley ST, Meinhart CD, Beebe DJ, Adrian RJ (1998) A particle image velocimetry system for microfluidics. Exp Fluids 25(4):316–319

Satyanarayana S, McCormick DT, Majumdar A (2006) Parylene micro membrane capacitive sensor array for chemical and biological sensing. Sens Actuators B 115(1):494–502

Sauer-Budge AF, Mirer P, Chatterjee A, Klapperich CM, Chargin D, Sharon A (2009) Low cost and manufacturable complete microTAS for detecting bacteria. Lab Chip 9(19):2803–2810

Scott A, Au AK, Vinckenbosch E, Folch A (2013) A microfluidic D-subminiature connector. Lab Chip 13:2036–2039

Schabmueller CGJ, Koch M, Mokhtari ME, Evans AGR, Brunnschweiler A, Sehr H (2002) Self-aligning gas/liquid micropump. J Micromech Microeng 12(4):420

Schönfeld F, Hessel V, Hofmann C (2004) An optimised split-and-recombine micro-mixer with uniform 'chaotic' mixing. Lab Chip 4(1):65–69

Schumacher S, Nestler J, Otto T, Wegener M, Ehrentreich-Förster E, Michel D, Wunderlich K, Palzer S, Sohn K, Weber A (2012) Highly-integrated lab-on-chip system for point-of-care multiparameter analysis. Lab Chip 12(3):464–473

Sen M, Wajerski D, Gad-el-Hak M (1996) A novel pump for MEMS applications. J Fluid Eng Trans ASME 118(3):624–627

Shikida M, Sato K, Tanaka S, Kawamura Y, Fujisaki Y (1994) Electrostatically driven gas valve with high conductance. J Microelectromech Syst 3(2):76–80

Sim WY, Yoon HJ, Jeong OC, Yang SS (2003) A phase-change type micropump with aluminum flap valves. J Micromech Microeng 13(2):286

Sin A, Reardon CF, Shuler ML (2004) A self-priming microfluidic diaphragm pump capable of recirculation fabricated by combining soft lithography and traditional machining. Biotechnol Bioeng 85(3):359–363

Singh MK, Anderson PD, Meijer HE (2009) Understanding and optimizing the SMX static mixer. Macromol Rapid Commun 30(4–5):362–376

Smithies O (1955) Zone electrophoresis in starch gels - group variations in the serum proteins of normal human adults. Biochem J 61(4):629–641

Smits JG (1985) Piezoelectric micropump for peristaltic fluid displacements. NL 8302860

Smits JG (1990) Piezoelectric micropump with three valves working peristaltically. Sensors Actuators A Phys 21(1–3):203–206

Song H, Bringer MR, Tice JD, Gerdts CJ, Ismagilov RF (2003) Experimental test of scaling of mixing by chaotic advection in droplets moving through microfluidic channels. Appl Phys Lett 83(22):4664–4666

Spencer WJ, Corbett WT, Dominguez LR, Shafer BD (1978) An electronically controlled piezo-electric insulin pump and valves. IEEE Trans Sonics Ultrasonics 25(3):153–156

Star A, Tu E, Niemann J, Gabriel J-CP, Joiner CS, Valcke C (2006) Label-free detection of DNA hybridization using carbon nanotube network field-effect transistors. Proc Natl Acad Sci U S A 103(4):921–926

Strohmeier O, Emperle A, Roth G, Mark D, Zengerle R, von Stetten F (2013) Centrifugal gas-phase transition magnetophoresis (GTM) - a generic method for automation of magnetic bead based assays on the centrifugal microfluidic platform and application to DNA purification. Lab Chip 13(1):146–155

Stroock AD, Dertinger SK, Ajdari A, Mezić I, Stone HA, Whitesides GM (2002) Chaotic mixer for microchannels. Science 295(5555):647–651

Studer V, Jameson R, Pellereau E, Pépin A, Chen Y (2004) A microfluidic mammalian cell sorter based on fluorescence detection. Microelectron Eng 73–74:852–857

Su Y-C, Lin L (2004) A water-powered micro drug delivery system. J Microelectromech Syst 13 (1):75–82

Sundararajan N, Kim D, Berlin AA (2005) Microfluidic operations using deformable polymer membranes fabricated by single layer soft lithography. Lab Chip 5(3):350–354

Suzuki H, Ho C-M (2002) A magnetic force driven chaotic micro-mixer. In: The fifteenth IEEE international conference on micro electro mechanical systems, 2002. IEEE, Washington, DC, pp 40–43
Suzuki H, Yoneyama R (2003) Integrated microfluidic system with electrochemically actuated on-chip pumps and valves. Sens Actuators B 96(1–2):38–45
Suzuki K, Fujiki I, Hagura Y (1998) Preparation of corn oil/water and water/corn oil emulsions using PTFE membranes. Food Sci Tech Int Tokyo 4(2):164–167
Takagi H, Maeda R, Ozaki K, Parameswaran M, Mehta M (1994) Phase transformation type micro pump. In: Proceedings, 5th international symposium on micro machine and human science, 1994. IEEE, Washington, DC, p 199
Takao H, Miyamura K, Ebi H, Ashiki M, Sawada K, Ishida M (2005) A MEMS microvalve with PDMS diaphragm and two-chamber configuration of thermo-pneumatic actuator for integrated blood test system on silicon. Sensors Actuators A Phys 119(2):468–475
Tan F, Leung PHM, Z-b L, Zhang Y, Xiao L, Ye W, Zhang X, Yi L, Yang M (2011) A PDMS microfluidic impedance immunosensor for E. coli O157:H7 and Staphylococcus aureus detection via antibody-immobilized nanoporous membrane. Sens Actuators B 159(1):328–335
Tas N, Berenschot J, Lammerink T, Elwenspoek M, Van den Berg A (2002) Nanofluidic bubble pump using surface tension directed gas injection. Anal Chem 74(9):2224–2227
Teh S-Y, Lin R, Hung L-H, Lee AP (2008) Droplet microfluidics. Lab Chip 8(2):198–220
Terray A, Oakey J, Marr DW (2002) Microfluidic control using colloidal devices. Science 296 (5574):1841–1844
Terry SC, Jerman JH, Angell JB (1979) Gas-chromatographic air analyzer fabricated on a silicon-wafer. IEEE Trans Electron Devices 26(12):1880–1886
Teymoori MM, Abbaspour-Sani E (2005) Design and simulation of a novel electrostatic peristaltic micromachined pump for drug delivery applications. Sensors Actuators A Phys 117(2):222–229
Thomas L Jr, Bessman S (1975) Prototype for an implantable micropump powdered by piezo-electric disk benders. Trans Am Soc Artif Int Organs 21:516
Thorsen T, Maerkl SJ, Quake SR (2002) Microfluidic large-scale integration. Science 298 (5593):580–584
Thorsen T, Roberts RW, Arnold FH, Quake SR (2001) Dynamic pattern formation in a vesicle-generating microfluidic device. Phys Rev Lett 86(18):4163–4166
Tice JD, Lyon AD, Ismagilov RF (2004) Effects of viscosity on droplet formation and mixing in microfluidic channels. Anal Chim Acta 507(1):73–77
Tice JD, Song H, Lyon AD, Ismagilov RF (2003) Formation of droplets and mixing in multiphase microfluidics at low values of the reynolds and the capillary numbers. Langmuir 19(22):9127–9133
Tiensuu A-L, Öhman O, Lundbladh L, Larsson O (2000) Hydrophobic valves by ink-jet printing on plastic CDs with integrated microfluidics. In: Berg A, Olthuis W, Bergveld P (eds) Micro total analysis systems. Springer, Dordrecht, The Netherlands, pp 575–578. doi:10.1007/978-94-017-2264-3_135
Tovar AR, Lee AP (2009) Lateral cavity acoustic transducer. Lab Chip 9(1):41–43
Truckenmüller R, Rummler Z, Schaller T, Schomburg K (2002) Low-cost thermoforming of micro fluidic analysis chips. J Micromech Microeng 12(4):375–379
Tsai J-H, Lin L (2002) Active microfluidic mixer and gas bubble filter driven by thermal bubble micropump. Sensors Actuators A Phys 97:665–671
Tsai JH, Liwei L (2002) A thermal-bubble-actuated micronozzle-diffuser pump. J Microelectromech Syst 11(6):665–671
Tsai R-T, Wu C-Y (2011) An efficient micromixer based on multidirectional vortices due to baffles and channel curvature. Biomicrofluidics 5(1):014103
Tsao T, Moroney R, Martin B, White R (1991) Electrochemical detection of localized mixing produced by ultrasonic flexural waves. In: Ultrasonics symposium, 1991, Proceedings. IEEE, Washington, DC, pp 937–940

Umbanhowar PB, Prasad V, Weitz DA (1999) Monodisperse emulsion generation via drop break off in a coflowing stream. Langmuir 16(2):347–351

Unger MA, Chou HP, Thorsen T, Scherer A, Quake SR (2000) Monolithic microfabricated valves and pumps by multilayer soft lithography. Science 288(5463):113–116

Van de Pol FCM, Van Lintel HTG, Elwenspoek M, Fluitman JHJ (1990) A thermopneumatic micropump based on micro-engineering techniques. Sensors Actuators A Phys 21(1–3):198–202

van der Wijngaart W, Ask H, Enoksson P, Stemme G (2002) A high-stroke, high-pressure electrostatic actuator for valve applications. Sensors Actuators A Phys 100(2–3):264–271

van Kan JA, Zhang C, Perumal Malar P, van der Maarel JRC (2012) High throughput fabrication of disposable nanofluidic lab-on-chip devices for single molecule studies. Biomicrofluidics 6 (3):36502

van Lintel HTG, van De Pol FCM, Bouwstra S (1988) A piezoelectric micropump based on micromachining of silicon. Sensors Actuators 15(2):153–167

van Oudheusden BW (1992) Silicon thermal flow sensors. Sensors Actuators A Phys 30(1–2):5–26

Veenstra T, Lammerink T, Elwenspoek M, Van Den Berg A (1999) Characterization method for a new diffusion mixer applicable in micro flow injection analysis systems. J Micromech Microeng 9(2):199

Vella SJ, Beattie P, Cademartiri R, Laromaine A, Martinez AW, Phillips ST, Mirica KA, Whitesides GM (2012) Measuring markers of liver function using a micropatterned paper device designed for blood from a fingerstick. Anal Chem 84(6):2883–2891

Vrouwe EX, Luttge R, van den Berg A (2004) Direct measurement of lithium in whole blood using microchip capillary electrophoresis with integrated conductivity detection. Electrophoresis 25 (10–11):1660–1667

Wagner B, Quenzer HJ, Hoerschelmann S, Lisec T, Juerss M (1996) Bistable microvalve with pneumatically coupled membranes. In: The ninth annual international workshop on micro electro mechanical systems (MEMS 1996), 11–15 Feb 1996. IEEE, Washington, DC, pp 384–388. doi:10.1109/MEMSYS.1996.494012

Waibel G, Kohnle J, Cernosa R, Storz M, Schmitt M, Ernst H, Sandmaier H, Zengerle R, Strobelt T (2003) Highly integrated autonomous microdosage system. Sensors Actuators A Phys 103 (1–2):225–230

Waldbaur A, Carneiro B, Hettich P, Wilhelm E, Rapp BE (2013a) Computer-aided microfluidics (CAMF): from digital 3D-CAD models to physical structures within a day. Microfluid Nanofluid 15(5):625–635

Waldbaur A, Kittelmann J, Radtke CP, Hubbuch J, Rapp BE (2013b) Microfluidics on liquid handling stations (μF-on-LHS): an industry compatible chip interface between microfluidics and automated liquid handling stations. Lab Chip 13(12):2337–2343

Waldbaur A, Rapp H, Länge K, Rapp BE (2011) Let there be chip – towards rapid prototyping of microfluidic devices: one-step manufacturing processes (cover article). Anal Methods 3 (12):2681–2716

Waldbaur A, Waterkotte B, Schmitz K, Rapp BE (2012) Maskless projection lithography for the fast and flexible generation of grayscale protein patterns. Small 8(10):1570–1578

Walker G, Ozers M, Beebe D (2004) Cell infection within a microfluidic device using virus gradients. Sens Actuators B 98(2):347–355

Walker GM, Beebe DJ (2002) A passive pumping method for microfluidic devices. Lab Chip 2 (3):131–134

Wang H, Chen Y, Hassibi A, Scherer A, Hajimiri A (2009) A frequency-shift CMOS magnetic biosensor array with single-bead sensitivity and no external magnet. In: IEEE international, 2009, solid-state circuits conference-digest of technical papers, 2009, ISSCC 2009. IEEE, Washington, DC, pp 438–439

Wang H, Iovenitti P, Harvey E, Masood S (2002) Optimizing layout of obstacles for enhanced mixing in microchannels. Smart Mater Struct 11(5):662

Wang Y-C, Choi MH, Han J (2004) Two-dimensional protein separation with advanced sample and buffer isolation using microfluidic valves. Anal Chem 76(15):4426–4431

Wang Y, Zhe J, Chung BT, Dutta P (2008) A rapid magnetic particle driven micromixer. Microfluid Nanofluid 4(5):375–389

Wego A, Pagel L (2001) A self-filling micropump based on PCB technology. Sensors Actuators A Phys 88(3):220–226

Weigl BH, Kriebel J, Mayes KJ, Bui T, Yager P (1999) Whole blood diagnostics in standard gravity and microgravity by use of microfluidic structures (T-sensors). Microchim Acta 131 (1–2):75–83

Wen CY, Yeh CP, Tsai CH, Fu LM (2009) Rapid magnetic microfluidic mixer utilizing AC electromagnetic field. Electrophoresis 30(24):4179–4186

Widmer HM (1983) Trends in industrial analytical-chemistry. Trac Trends Anal Chem 2(1):R8–R10

Wilhelm E, Neumann C, Duttenhofer T, Pires L, Rapp BE (2013a) Connecting microfluidic chips using a chemically inert, reversible, multichannel chip-to-world-interface. Lab Chip 13 (22):4343–4351

Wilhelm E, Neumann C, Sachsenheimer K, Schmitt T, Lange K, Rapp BE (2013b) Rapid bonding of polydimethylsiloxane to stereolithographically manufactured epoxy components using a photogenerated intermediary layer. Lab Chip 13(12):2268–2271

Winkley J, Yanowski L, Hynes W (1937) A systematic semimicro procedure for the qualitative analysis of the commoner cations. Mikrochemie 21(1):102–115

Witek MA, Llopis SD, Wheatley A, McCarley RL, Soper SA (2006) Purification and preconcentration of genomic DNA from whole cell lysates using photoactivated polycarbonate (PPC) microfluidic chips. Nucleic Acids Res 34(10):e74

Woias P, Hauser K, Yacoub-George E (2000) An active silicon micromixer for μTAS applications. In: Berg A, Olthuis W, Bergveld P (eds) Micro total analysis systems. Springer, Dordrecht, The Netherlands, pp 277–282

Wong SH, Ward MC, Wharton CW (2004) Micro T-mixer as a rapid mixing micromixer. Sens Actuators B 100(3):359–379

Worgull M, Kolew A, Heilig M, Schneider M, Dinglreiter H, Rapp BE (2011) Hot embossing of high performance polymers. Microsyst Technol 17(4):585–592

Wu C-Y, Liao W-H, Tung Y-C (2011) Integrated ionic liquid-based electrofluidic circuits for pressure sensing within polydimethylsiloxane microfluidic systems. Lab Chip 11(10):1740–1746

Xing Y, Grosjean C, Yu-Chong T (1999) Design, fabrication, and testing of micromachined silicone rubber membrane valves. J Microelectromech Syst 8(4):393–402

Yamahata C, Lacharme F, Burri Y, Gijs MAM (2005) A ball valve micropump in glass fabricated by powder blasting. Sens Actuators B 110(1):1–7

Yan D, Yang C, Miao J, Lam Y, Huang X (2009) Enhancement of electrokinetically driven microfluidic T-mixer using frequency modulated electric field and channel geometry effects. Electrophoresis 30(18):3144–3152

Yang B, Lin Q (2009) A latchable phase-change microvalve with integrated heaters. J Microelectromech Syst 18(4):860–867

Yang J, Liu Y, Rauch CB, Stevens RL, Liu RH, Lenigk R, Grodzinski P (2002) High sensitivity PCR assay in plastic micro reactors. Lab Chip 2(4):179–187

Yang X, Grosjean C, Tai Y-C, Ho C-M (1998) A MEMS thermopneumatic silicone rubber membrane valve. Sensors Actuators A Phys 64(1):101–108

Yang Z, Goto H, Matsumoto M, Maeda R (2000) Active micromixer for microfluidic systems using lead-zirconate-titanate(PZT)-generated ultrasonic vibration. Electrophoresis 21(1):116–119

Yaralioglu GG, Wygant IO, Marentis TC, Khuri-Yakub BT (2004) Ultrasonic mixing in microfluidic channels using integrated transducers. Anal Chem 76(13):3694–3698

Yardley SJS, Linkenheimer WH (1971) Osmotic fluid reservoir for osmotically activated long-term continuous injector device. United States Patent

Yasuda K (2000) Non-destructive, non-contact handling method for biomaterials in micro-chamber by ultrasound. Sens Actuators B 64(1):128–135

Ymeti A, Greve J, Lambeck PV, Wink T, Stephan WFM, Tom AM, Wijn RR, Heideman RG, Subramaniam V, Kanger JS (2006) Fast, Ultrasensitive Virus Detection Using a Young Interferometer Sensor. Nano Lett 7(2):394–397

Yoshida K, Kikuchi M, Park JH, Yokota S (2002) Fabrication of micro electro-rheological valves (ER valves) by micromachining and experiments. Sensors Actuators A Phys 95(2–3):227–233

Young EW, Berthier E, Guckenberger DJ, Sackmann E, Lamers C, Meyvantsson I, Huttenlocher A, Beebe DJ (2011) Rapid prototyping of arrayed microfluidic systems in polystyrene for cell-based assays. Anal Chem 83(4):1408–1417

Yu-Chuan S, Liwei L, Pisano AP (2002) A water-powered osmotic microactuator. J Microelectromech Syst 11(6):736–742

Zeng J, Banerjee D, Deshpande M, Gilbert JR, Duffy DC, Kellogg GJ (2000) Design analyses of capillary burst valves in centrifugal microfluidics. In: Proceedings of the micro total analysis systems symposium (lTAS 2000) May, 2000. pp 14–18

Zeng S, Chen C-H, Mikkelsen JC Jr, Santiago JG (2001) Fabrication and characterization of electroosmotic micropumps. Sens Actuators B 79(2–3):107–114

Zeng S, Chen C-H, Santiago JG, Chen J-R, Zare RN, Tripp JA, Svec F, Fréchet JM (2002) Electroosmotic flow pumps with polymer frits. Sens Actuators B 82(2):209–212

Zengerle R, Richter A, Sandmaier H (1992) A micro membrane pump with electrostatic actuation. In: Micro electro mechanical systems, 1992, MEMS'92, Proceedings An investigation of micro structures, sensors, actuators, machines and robot, 4–7 Feb 1992. IEEE, Washington, DC, pp 19–24. doi:10.1109/memsys.1992.187684

Zengerle R, Richter M (1994) Simulation of microfluid systems. J Micromech Microeng 4(4):192

Zengerle R, Ulrich J, Kluge S, Richter M, Richter A (1995) A bidirectional silicon micropump. Sensors Actuators A Phys 50(1–2):81–86

Zhao B, Moore JS, Beebe DJ (2001) Surface-directed liquid flow inside microchannels. Science 291(5506):1023–1026

Chapter 6
Microfluidic Electrochemical Biosensors: Fabrication and Applications

Sandrine Miserere and Arben Merkoçi

Abstract A biosensor is generally defined as an analytical device which converts a biological response into a quantifiable and processable signal (Christopher R. Lowe 1984). It consists of bio-recognition systems (enzymes, DNA, proteins, cells, etc.), immobilized onto the surface of a transducer (electrochemical, optical, mass and thermal for the most common). Specific interactions between the target (analyte) and the complementary bio-recognition layer produce a physico-chemical change, which is detected and may be measured by the transducer. This specific interaction leads to very selective and specific sensors.

When an electrode constitutes the transducer, we will talk about electrochemical biosensors, which are capable to transduce electrochemical reactions. Electrochemical biosensors are reliable, selective and ultra-low detection can be achieved (Grieshaber et al. 2008).

Biosensors are particularly used for pathogen detection, molecular diagnostics, environmental monitoring, food safety control as well as in defense in order to produce point of care devices, but in many cases the main limitation in realizing such kind of devices is the ability to miniaturize the transduction principle and the lack of a cost-effective production method. That is why electrochemical biosensors compared to other detection seem to be better candidates. Indeed electrical signals can be processed by conventional electronics in a very cheap and fast manner. The integration and miniaturization of electrochemical transducers in a microsystem make easier its fabrication in front of the most common currently used detection methods as fluorescence-based (Grieshaber et al. 2008).

Over the last years, the use of microfluidic platforms for electroanalysis has increased in a remarkable way. This can be explained by the fact that microdevices

S. Miserere • A. Merkoçi (✉)
Nanobioelectronics & Biosensors Group, Catalan Institute of Nanosciences and Nanotechnology (ICN2), Autonomous University of Barcelona, 08193, Bellaterra, Barcelona, Spain

Catalan Institution for Research and Advanced Studies (ICREA), 0801 Barcelona, Spain
e-mail: arben.merkoci@icn.cat

J. Castillo-León, W.E. Svendsen (eds.), *Lab-on-a-Chip Devices and Micro-Total Analysis Systems*, DOI 10.1007/978-3-319-08687-3_6

work under hydrodynamic conditions which enhance mass transport toward the detector surface resulting in an increase of the obtained electrical current and sensitivity compared to the classical static measurement modes. Furthermore, simple and miniaturized microplatforms are especially interesting for their low consumption of sample and reagent, smaller times of analysis, increased reliability and repeatability, portability and the integration of conventional analytical techniques.

Considering integration capability, electrochemistry seems to be a good candidate as detection part for its versatility of size, geometry and nature of electrodes that can be integrated within a microfluidic platform, and the minimum instrumentation it requires. Moreover as mentioned before, electrochemical detection is reliable, selective and ultra-low detection can be achieved.

Combined together, electrochemistry and microfluidic could provide a really powerful biosensor platform.

6.1 Electrodes Fabrication

Integration of electrodes onto plastic and glass chips can greatly enhance the functionality and automation. Integrated electrodes also can help to minimize contamination for high-throughput screening and point of care (POC) devices compared with external, reusable wire electrodes.

The electrodes inside fluidic channels are used, for example, for electrical impedance spectrometry, contact or contactless conductivity detection, amperometric detection, electrical heating and temperature sensing. Electrodes inside fluidic compartments are mainly used to apply high voltages to generate electroosmotic flow or to perform electrophoresis.

Electrodes can be integrated between two glass or silicon layers or placed on top of the chip. They can be also fabricated onto plastic substrates.

In the following section, different methods to deposit electrodes will be exposed.

6.1.1 Sputtering and Evaporation

Thin film electrodes depositions techniques are the most classical techniques to fabricate electrodes for microfluidics applications. Depending on the electrode material sputtering or evaporation can be used.

To sputter conducting targets, a DC power supply is generally used. In a DC diode sputtering system, as it can be seen in Fig. 6.1, argon is ionized by a strong potential difference and these ions are thus accelerated to the surface of the target. The impact causes the release of atoms of the target material which travel to the substrate, condense and form a thin layer. This process is an isotropic process where sputtered atoms ejected into the gas phase tend to deposit on all surfaces in the vacuum chamber.

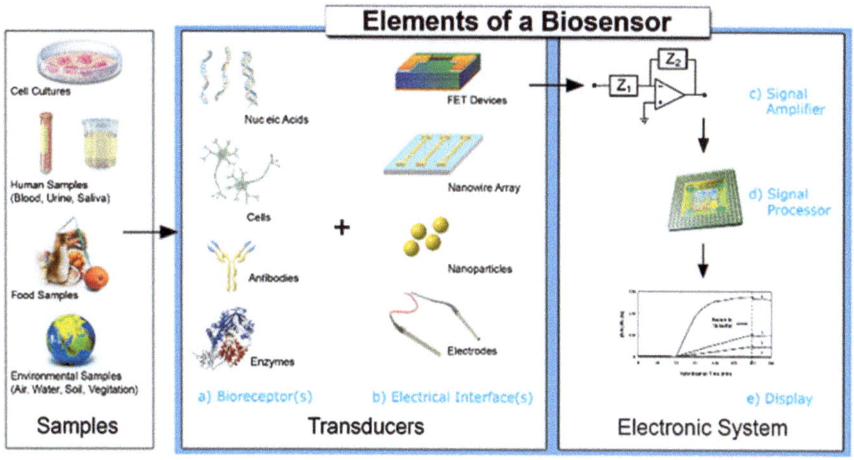

Fig. 6.1 Biosensor principle. Reprinted from (Grieshaber et al. 2008) Copyright 2008 MDPI

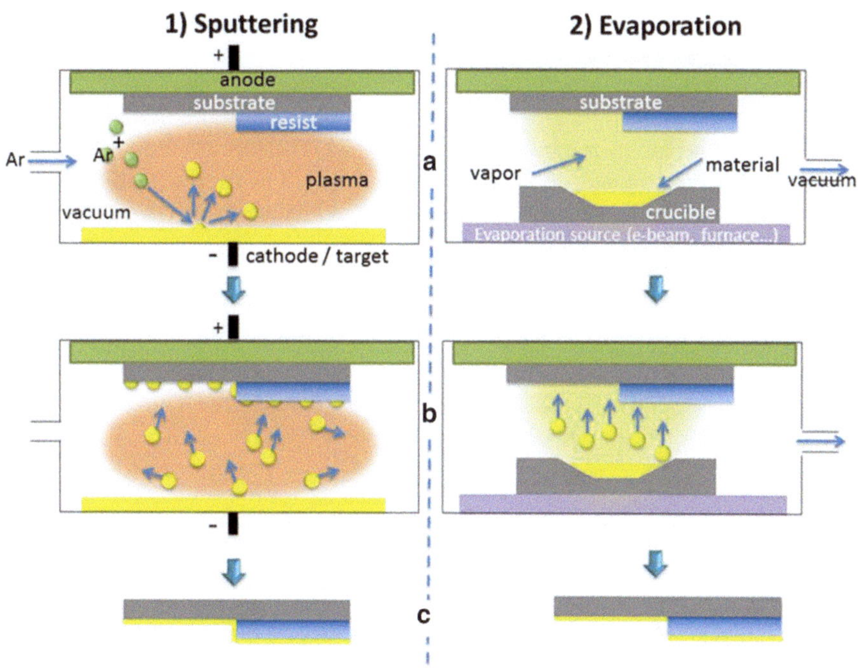

Fig. 6.2 Thin film deposition techniques: (1) sputtering process; (2) evaporation process

In the case of evaporation, the material to be deposited is evaporated under vacuum conditions. This evaporation could be performed by a beam of electrons (e-beam evaporation (Bowden et al. 1998)) as it can be seen in Fig. 6.2, chemically (chemical vapor deposition, CVD (Crosby and Hanley 1977)) or by a hot filament (physical vapor deposition, PVD (Reichelt and Jiang 1990)). When the material enters in contact with the substrate, it condenses and comes back to its solid state, forming our thin layer. This process in contrast to sputtering is an anisotropic process which makes evaporation a better candidate when lift-off process is used to fabricate electrodes as it will be explained in the following sections. Temperature achieved with evaporation could be really high which limits the type of the substrates. Indeed polymer substrates with low glass transition temperature could not be used. For example, in the case of carbon or boron doped diamond electrodes, only silicon wafer could be used as the evaporation is performed at 1,000 ºC (Peckova et al. 2009).

In both cases, electrodes fabrication protocol is quite the same as shown in Fig. 6.3. A substrate is spin-coated with a positive resist and then exposed to UV through a mask. The exposed parts are developed thanks to a solvent and let free the substrate to be covered by the electrode material which is deposited through evaporation or sputtering as explain above. Afterwards, a lift-off step (Fig. 6.3) is performed in order to strip the underlying resist and reveal the geometry of the electrodes.

Etching process can also be used (Fig. 6.3). In this case, a thin film is deposited onto the substrate and then spin-coating of the resist is performed and succeeded by the photolithography. Finally, the material not recovered by the resist will be etched and the geometry of the electrode will appear. In a final step, the resist which stay onto the electrodes will be stripped.

When gold is deposited, it can be useful to deposit a previous layer such as titanium or chrome to increase the mechanical resistance of the electrode.

Although these techniques achieve really high resolution, they are limited by the requirement of special and expensive equipment. Most of the time, they are done in clean-room environment. That is why alternative methods in order to produce low cost and high throughput devices tend to be developed.

6.1.2 *Inkjet Printing*

Inkjet printing technology is a direct writing technique used for office printing and which becomes a serious candidate to produce flexible and low-cost electronics. It is frequently used for the fabrication of organic electronic devices and tends also to be used to produce complete electrochemical biosensors (Komuro et al. 2013; Maattanen et al. 2013).

The advantages of inkjet technology are its low cost, simplicity, high resolution, speed, reproducibility, flexibility, noncontact, and low amount of waste generated. A huge variety of substrates (glass, plastics, ceramics, and paper) and inks can be

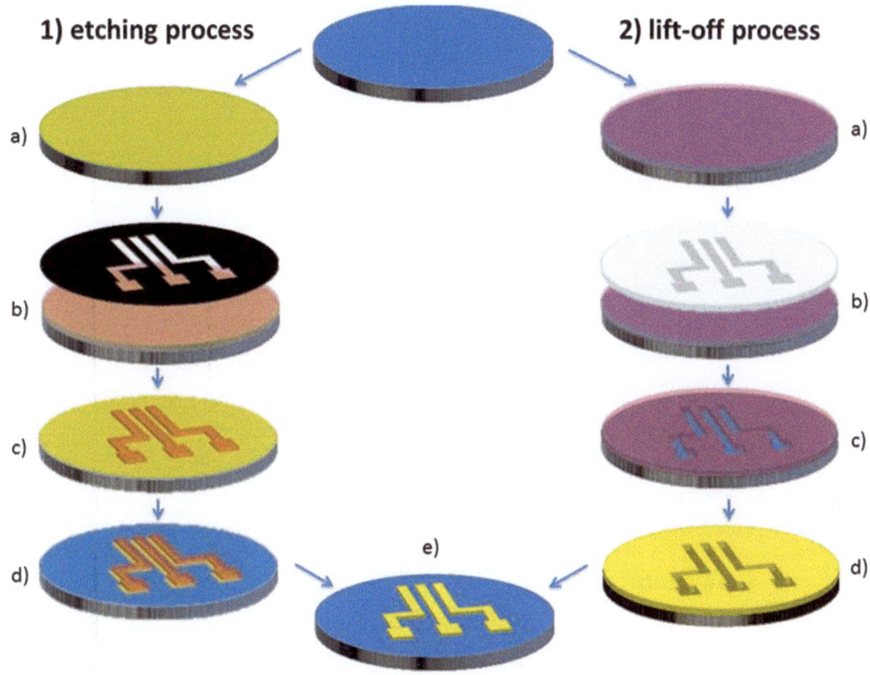

Fig. 6.3 Electrodes fabrication through etching (1) or lift-off (2) processes

used. Different kind of inks can be deposited by inkjet from metallic nanoparticles solutions (silver (Perelaer et al. 2008), copper (Li et al. 2009), gold (Jensen et al. 2011) to carbon based solutions (carbon nanotubes (Tortorich and Choi 2013), graphene (Li et al. 2013) and conducting polymer (PEDOT (Xiong and Liu 2012).

Figure 6.4 shows how inkjet printing works. In summary, a drop of ink is ejected from a nozzle by thermal or piezo actuation and falls onto a substrate creating a design. Depending of the ink/substrate interactions, reliable direct patterning with line and space dimensions in the 10–100 μm range can be produced. The thickness of these electrodes is in the range of hundreds of nanometer.

6.1.3 Screen Printing

Screen printing is a printing technique in which a design is imposed on a screen of polyester or other fine mesh, with blank areas coated with an impermeable substance. Ink is forced into the mesh by the squeegee and onto the surface to be printed. By using several screens, different layers can be printed. Nowadays, screen-printing microfabrication technology is a well-established technology for mass production of thick film electrodes and is widely used to produce low cost

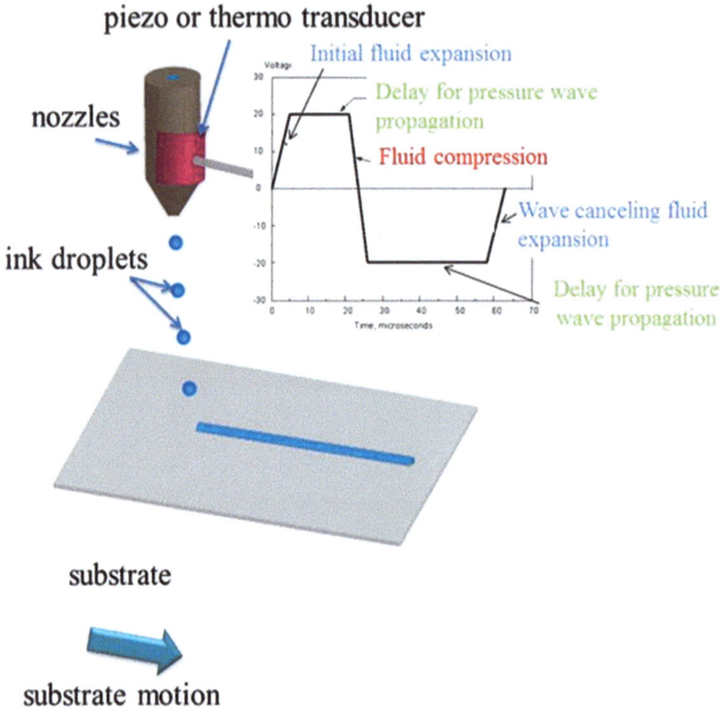

Fig. 6.4 Inkjet printing principle

sensors. Several companies commercialized these types of electrodes such as DropSens (http://www.dropsens.com/) or BioLogic (http://www.bio-logic.info/).

The protocol to print three electrodes electrochemical cell constituted by carbon and silver/silver chloride electrodes is explained in Fig. 6.5. A first mask is used to print the carbon layer (working and counter electrode). The substrate could be a plastic one (generally polyethylene terephtalate, PET), ceramic, or paper. These electrodes are cured at temperature given by the provider of the ink. Then by using a second mask, the reference electrode is printed by using silver/silver chloride ink. A final insulator layer can be printed but in the case of microfluidics the area of contact of the electrodes will be delimited by the microchannel itself.

6.1.4 Reference Electrodes

A reference electrode (RE) is an electrode with a fixed and known potential in order to measure the potential of an electrochemical interface. This potential should be insensitive to any changes occurring in the electrolyte solution. This is one of the

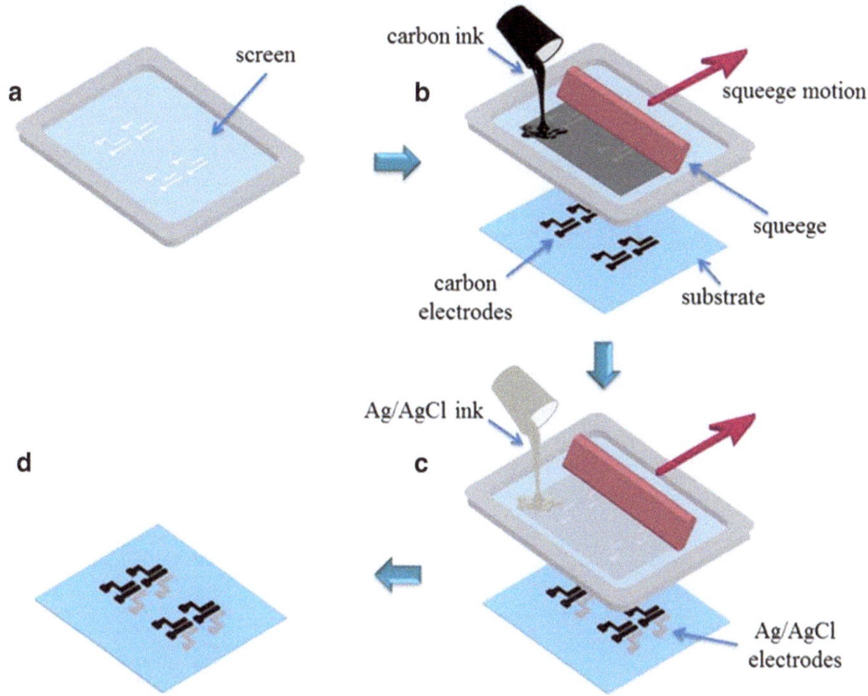

Fig. 6.5 Screen printing of three electrodes electrochemical cell (**a**) screen, (**b**) carbon layer deposition, (**c**) Ag/AgCl deposition, (**d**) final electrochemical cell

most important parts of an electrochemical sensor and the accuracy of the measurements strongly depends on the quality of this electrode (Rius-Ruiz et al. 2011).

Ag/AgCl electrodes are the most common electrodes used as RE. In order to obtain such kind of electrodes, a film of silver is deposited by the different techniques already described in this chapter (sputtering, evaporation, inkjet, etc.) and the surface chemically or electrochemically modified to form a thin silver chloride film.

In the case of screen printing, no additional step is required as the ink used to print the reference electrode is constituted by a mix of silver/silver chloride. The produced electrode will be considered as a pseudo-reference.

It is important to keep in mind that this type of electrodes remains pseudo-reference electrodes as the stability of their potential is related to the stability of the AgCl film. Indeed this one tends to be dissolved and thus changes dramatically the potential of the electrode. In order to prevent this dissolution a protective layer such as Nafion (Moussy and Harrison 1994) can be added. Another solution consists in introducing a junction with a saturated solution of chlorine ions (Lauks 1989).

Gold, platinum, or even carbon has also been used depending on the nature of the analyte (Mehrvar and Abdi 2004).

6.2 Electrochemical Detection Techniques

Typically in electrochemistry, the studied reaction will produce a measurable current (*amperometry*), a measurable potential (*potentiometry*) or measurably change of conductivity of the solution (*conductometry*) between electrodes (Chaubey and Malhotra 2002). Other types of electrochemical detection techniques exist such as *impedimetric* (*which measures* both resistance and reactance) *or field effect* (which uses transistor technology to measure current as a result of a potentiometric effect at a gate electrode).

We will introduce these techniques in the following part in order to have a better comprehension in the application part.

6.2.1 Amperometric Devices

Amperometric devices measure the current generated by an analyte, which reacts onto the electrode surface. A basic system of three-electrodes is generally used. It consists of a working electrode, a reference electrode, and a counter electrode. A polarizable electrode is used as working electrode to analyze electrode reactions. The reference electrode (as we have seen in Sect. 6.2.4) works as a potential standard, and a non-polarizable electrode is preferred. An appropriate potential is applied to the working electrode with respect to the reference electrode and the generated current between the working electrode and the auxiliary electrode is measured. Miniaturization and integration of amperometric sensors are easily achieved by creating a three-electrode system in the form of thin- or thick-film patterns (Grieshaber et al. 2008). Amperometric detection is usually used as detection in electrophoresis microchips (Fischer et al. 2009).

6.2.2 Potentiometric Devices

Potentiometric devices measure the accumulation of a charge potential at the working electrode compared to the reference electrode when zero or no significant current flows (conditions of equilibrium) between them. For potentiometric measurements, concentration and the potential are related and given by the Nernst equation (6.1).

$$E = E_0 + (RT/nF)\ln Q \tag{6.1}$$

where E represents the observed potential at zero current, E_o is a constant potential contribution to the cell, R is the universal gas constant, T is the absolute temperature in degrees Kelvin, n is the charge number of the electrode reaction, F is the Faraday constant, and Q is the ratio of ion concentration at the anode to ion concentration at the cathode (Grieshaber et al. 2008).

Potentiometric sensors are suitable for measuring low concentrations in tiny sample volumes. Many potentiometric devices are also based on various forms of field-effect transistor (FET) devices to measure pH changes (Ferrigno et al. 2004).

6.2.3 Conductimetric Devices

Conductimetric devices measure the ability of an analyte to conduct an electrical current between two electrodes. The conductimetric transducer is a miniature two-electrode device designed to measure the conductivity of the thin electrolyte layer adjacent to the electrode surface. One of the better configurations for conductimetric measurement is interdigitated structure (Sheppard et al. 1993).

Biosensors based on the conductimetric principle present a number of advantages. Thin-film electrodes are suitable for miniaturization and large scale production using inexpensive technology. They do not require any reference electrode, and the voltage can be sufficiently low to decrease the power consumption. Furthermore, a huge kind of compounds can be detected on the basis of various reactions and mechanisms.

In the most cases conductimetric devices have been strongly associated with enzymes as an enzyme reaction generates a variation of the ionic strength, and thus on the conductivity (Marrakchi et al. 2005; Soldatkin et al. 2012).

6.2.4 Impedimetric Devices

Impedimetric devices can study material property or specific processes that will change the conductivity/resistivity or capacitivity on the surface of an electrochemical system. They can monitor changes in electrical properties arising from biorecognition events at the surfaces of modified electrodes. For example, changes in the conductance of the electrode can be measured as a result of protein immobilization and antibody–antigen reactions on the electrode surface (Guan et al. 2004; Kell and Davey 1990).

6.3 Applications

6.3.1 Enzymatic-Based LOC Biosensors

Enzymes are proteins that act as biocatalysts to convert substrates into products. In some cases, enzymes catalyze reactions that produce or consume electrons and in this case these are also known as redox enzymes. The substrate is recognized by a part of the enzyme, similar to the antibody–antigen interaction, that is why these enzymes are used in biosensing. As a consequence of the current produced by this interaction, the concentration of the analyte can be calculated (Heller 1996).

The most commonly used enzymes in biosensing are glucose oxidase (GOx) (Fang et al. 2003) and horseradish peroxidase (HRP) (Azevedo et al. 2003).

Usually enzymes are immobilized onto the electrode surface through polymers such as polyethylene glycol, glutaraldehyde, or hydrogels (Brena et al. 2013).

Liu et al. (2004) (Fig. 6.6a) have proposed in 2004 the immobilization of DNA onto the interior walls of poly(dimethylsiloxane) (PDMS) microchannel and its application to an enzyme-amplified electrochemical DNA assay. DNA was immobilized through silanization of the PDMS surface and was used as capture probe. Then target DNA was introduced and finally modified with alkaline phosphatase (AP) conjugated to streptavidin.

Electrochemical detection of DNA was carried out using an indium tin oxide interdigitated array (IDA) electrode. The IDA electrode eliminates the need for a reference electrode and provides a steady-state current that is related to the concentration of hybridized DNA. A limit of detection of the DNA target of 1 nM in a sample volume of 20 nL, which corresponds to 20 attomoles of DNA was reached.

In 2011, Chen and White (2011) (Fig. 6.6b) have proposed an enzyme-linked assay on interdigitated electrodes. Seventy nanolitre sample volume was analyzed and the detection limit for 4-aminophenol (PAP) of 1.0×10^{-10} M was calculated. This low detection limit was due to the increased sensitivity provided by the sample confinement in the microfluidic channel, as well as the increased repeatability due to precisely controlled static flow . Furthermore, surface treatment was really efficient in order to decrease unspecific interactions.

Puri et al. (2013) (Fig. 6.6c) have developed a miniaturized electrochemical uric acid biosensor onto modified indium tin oxide (ITO) microelectrode array (μEA). The electrode was modified by uricase in order to detect uric acid through amperometric measurements in aqueous solutions. The sensor showed a linear response over a concentration range of 0.058–0.71 mM with a sensitivity of 46.26 μA/mM/cm^2. A response was obtained at 40 s reaching a 95 % steady-state current value. Furthermore, the enzymes remained active for 6 weeks.

Mayorga-Martinez et al. (2014) (Fig. 6.6d) have reported an integrated phenol sensoremoval microfluidic nanostructured platform. Screen-printed electrodes were integrated into a PDMS platform and modified by $CaCO_3$-poly(ethyleneimine) (PEI) nanostructured microparticles. These microparticles were themselves modified with tyrosinase. This device was capable to detect phenol compounds through

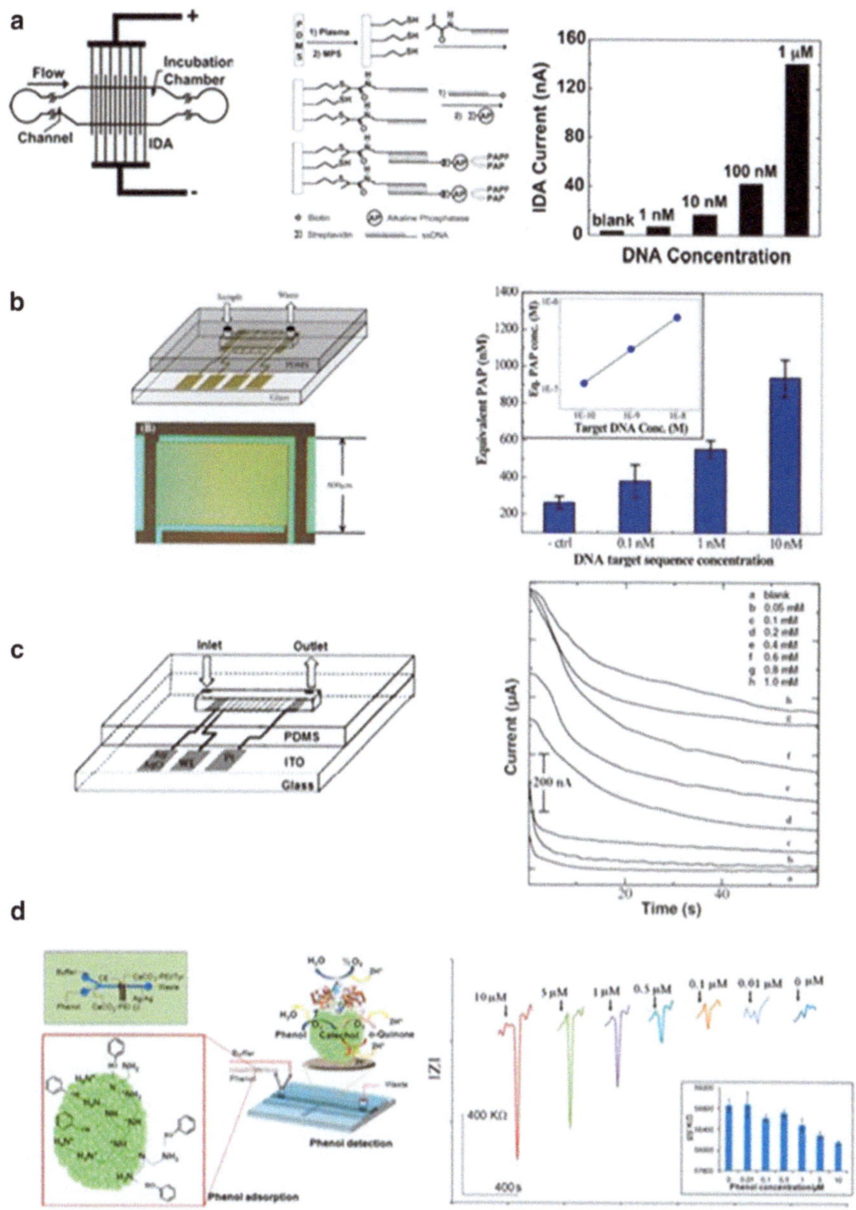

Fig. 6.6 Enzymes-based electrochemical LOC, (**a**) and (**b**) enzyme-amplified DNA hybridization assays (**a**) reprinted with permission from (Liu et al. 2004) Copyright 2004 American Chemical Society, (**b**) reprinted with permission from (Chen and White 2011) Copyright 2011 Elsevier, (**c**) reprinted with permission from (Puri et al. 2013) Copyright 2013 Springer (**d**), (**b**) reprinted with permission from (Mayorga-Martinez et al. 2014) Copyright 2014 Elsevier

chronoimpedimetric measurements and remove the same pollutant through its adsorption onto a nanostructurated polymer modified vaterite. Catechol as proof of concept compound was detected in a range of 0.01–10 mM achieving a limit of detection (LOD) of 4.64 nM.

6.3.2 DNA LOC Biosensors

An electrochemical detection system for DNA sensing can be achieved by the catalyzed reduction and oxidation of DNA bases or through the electrochemical response done by redox markers when a capture probe binds in a specific way with a capture probe.

To achieve such sensors, the surface of the electrode is modified by single strand DNA (ssDNA) through different chemistry (thiol, biotin-streptavidin, etc.). This ssDNA will serve as capture probe on the surface of the electrode. Once the capture has been immobilized, the blank can be measured. The target (or capture DNA) will be added and will change the signal provided by the electrode. The detection of the DNA can be improved by modification of the DNA with electroactive compounds (mediator like methylene blue (Malaquin et al. 2010), or metallic nanoparticles (AuNPs) (Lin et al. 2005), QDs (Marin and Merkoci 2009).

Using microfluidics, purification of the sample followed by its amplification (PCR, isothermal amplification) can be performed in the same platform, which leads as we have seen before to a reduction of the cost and duration of the analysis, reduction of the sample, and also errors that can occur by multi-manipulation of the sample.

Lin et al. have developed in 2011 (Lin et al. 2011) an organic electrochemical transistor (OECTs) with PEDOT:PSS as the active layer integrated into a flexible microfluidic system (Fig. 6.7a). The sensor is a label-free DNA sensor where ssDNA probes were immobilized onto the surface of the Au gate electrode. The device could successfully detect complementary DNA targets at concentrations as low as 1 nM and its detection limit is extended to 10 pM by pulse-enhanced hybridization of DNA.

In 2011, Ferguson et al. (2011) (Fig. 6.7b) have reported the use of a magnetic integrated microfluidic electrochemical detector (MIMED) that integrates sample preparation and electrochemical sensors to detect RNA-based virus directly from throat swab samples. They have used immunomagnetic target capture, concentration, and purification. Then they proceeded to reverse-transcriptase polymerase chain reaction (RT-PCR) to amplify the sample and finally detected the sample onto the surface of a gold electrode modified with a capture probe. They demonstrated the detection of influenza H1N1in throat swab samples at loads as low as ten TCID50, four orders of magnitude below the clinical titter for this virus.

In 2009, Swensen et al. (2009) (Fig. 6.7c) have shown a biosensor system capable of continuous and real-time measurement of cocaine. This device was based on the use of target-specific DNA aptamers. They generate an

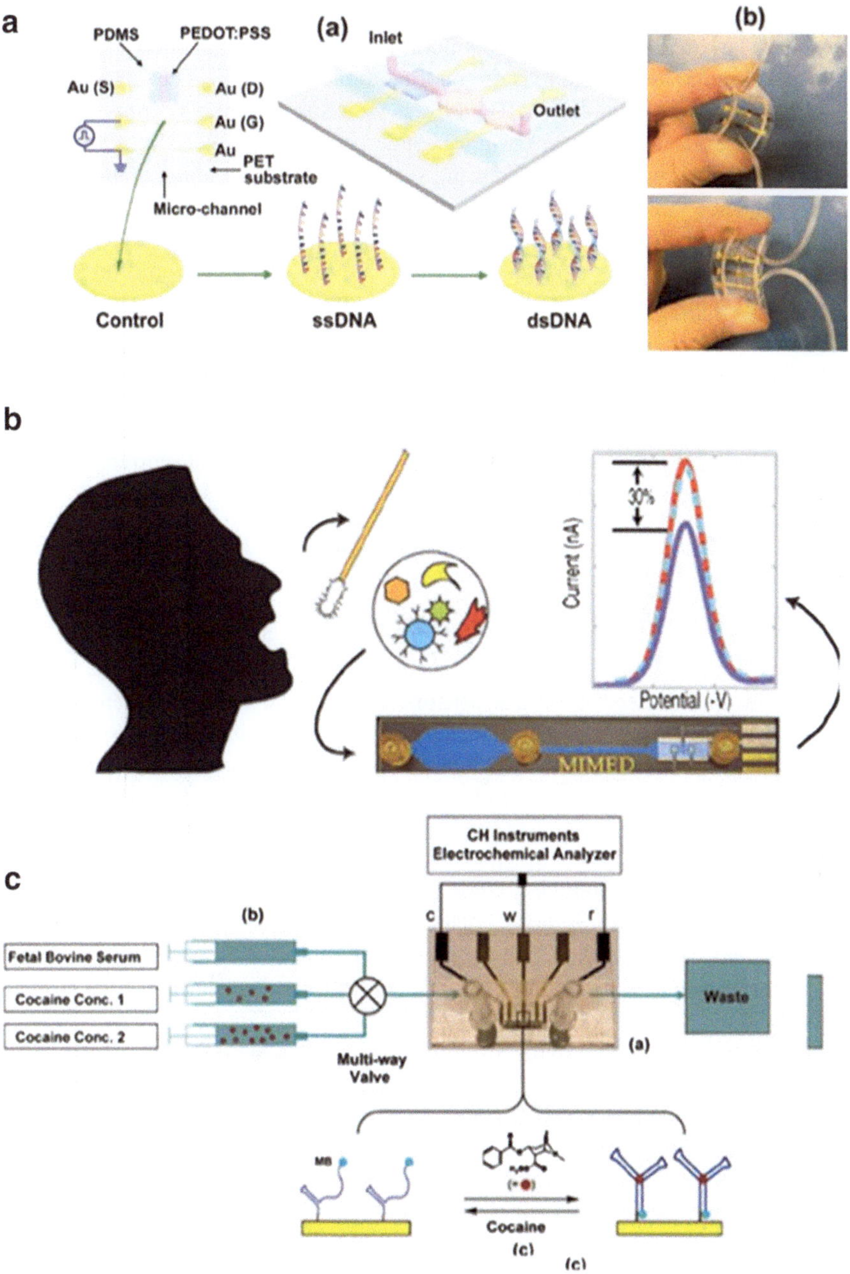

Fig. 6.7 DNA-based electrochemical LOC, (**A**) OECT integrated in a flexible microfluidic system, reprinted with permission from (Lin et al. 2011) Copyright 2011 Wiley, (**B**) MIMED biosensor reprinted with permission from (Ferguson et al. 2011) Copyright 2009 American Chemical Society, (**C**) aptamer biosensor for the detection of cocaine, reprinted with permission from (Swensen et al. 2009) Copyright 2009 American Chemical Society

electrochemical signal in response to the analyte with a microfluidic detection system. They demonstrated the continuous and real-time detection of cocaine in low micromolar concentrations in undiluted blood serum. The aptamers were modified with methylenblue (MB) particles which were far away from the surface of the electrode. When the cocaine bonded to the aptamer, its conformation changed, approaching the MB from the surface and thus creating an electrochemical signal.

In 2010, Malaquin et al. (2010) have reported the use of screen printed electrodes into polymeric microchip for real-time DNA quantification for low-cost neusocomial infections diagnosis. The system contained a DNA extraction zone (constituted by magnetic particles) followed by amplification. A redox compound was introduced together with DNA sample enabling electrochemical detection through square wave voltammetry (SQWV). This compound interacts with double-stranded DNA (dsDNA) by intercalation during the extension step, thus reducing its mobility and redox activity. As a result, the peak current decreases during amplification of the target DNA.

6.3.3 Antibodies-Based LOC-Biosensors

Antibodies-based electrochemical sensors, also known as immunosensors, offer sensitivity, selectivity and reliability at a low cost, making them very attractive tools for protein detection. Antibodies are most widely used because of the high specificity of the antibody–antigen binding.

After immobilizing antibodies onto a surface, the analyte or the so-called antigen is introduced and binds specifically to the capture antibody. In the most common cases, a secondary labeled antibody is then introduced and binds to the analyte in order to detect its concentration. This antibody can be labeled with metallic nanoparticles or enzymes. Label-free detection can also be performed generally by using impedance techniques which are sensitive to surface modification/changes.

Antibodies can be immobilized onto the surface of the channel as reported by Jang et al. (2006) (Fig. 6.8a). Antibodies were immobilized onto PDMS channel, chemically modified with a silane monolayer containing poly(ethylene glycol) and avidin linkage. PEG groups reduce the negative effect of non-specific adsorption, while avidin was used as a linker to capture the antibodies. A sandwich immunoassay was performed with alkaline-phosphatase (AP)-labeled secondary antibody to determine mouse IgG as a model analyte. A LOD (485 pg/mL) with 95 % confidence level was achieved.

Another option is to immobilize the antibodies directly onto the surface of the electrode. In 2014, Medina-Sanchez et al. (2014) have reported the immobilization of antibodies onto the surface of magnetic beads for the detection of ApoE, an eventual biomarker for Alzheimer disease. Cadmium-based quantum dots were

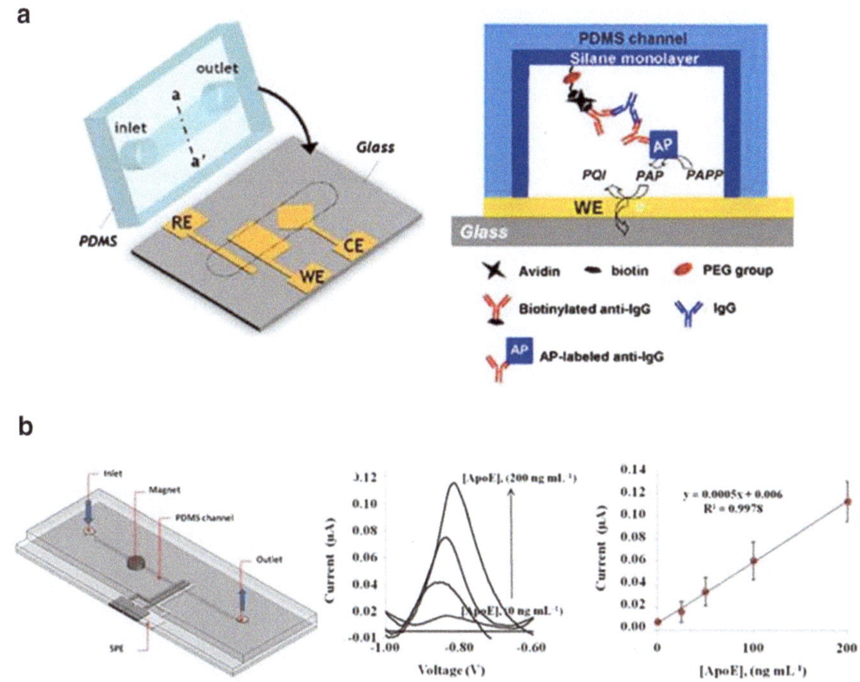

Fig. 6.8 Electrochemical immunosensing LOC. (**a**) Immunosensor for the detection of IgG, reprinted from (Jang et al. 2006) Copyright 2006 Elsevier; (**b**) LOC for the detection of ApOE Alzheimer biomarker, reprinted from (Medina-Sanchez et al. 2014) Copyright 2014 Elsevier

used as reporter and a LOD of 12.5 ng/ApoE/mL was achieved. Furthermore, serum samples were studied with a high correlation.

Kim et al. (2005) have reported a flow-enhanced electrochemical immunosensor on centrifugal microfluidic platforms. It was based on bead-based enzyme-linked immune-sorbent assay, and flow-enhanced electrochemical detection. They achieved a complete and automated assay in less than 20 min with a limit of detection (LOD) of 4.9 pg mL^{-1} for the C-reactive protein (CRP). This LOD was 17 times higher than the one obtained by optical method.

6.3.4 Cell-Based LOC-Biosensors

In the case of cell-based biosensor, living cells constitute the recognition element. They are high sensitive, selective and offer rapid response and versatility. They help to understand cellular mechanisms of particular diseases to improve the development of targeted treatment and can be applied in *food safety, environmental monitoring, drugs screening*, and *medical diagnostics*.

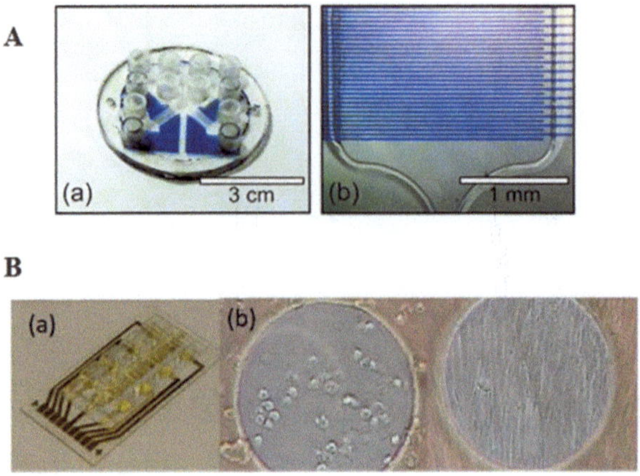

Fig. 6.9 Cell-based electrochemical LOC. (**a**) Polymer-based biosensor for rapid electrochemical detection of virus infection of human cells, reprinted with permission from (Kiilerich-Pedersen et al. 2011) Copyright 2011 Elsevier, (**b**) impedance biosensor showing an eight gold electrodes array (**a**) with cells before (*left*) and after (*right*) attachment to the gold electrode (**b**) reprinted from (Hondroulis et al. 2010) Copyright 2010 IOP

Microfluidics is convenient when cells are studied because the physiological and electrical responses of even single cells can be detected. Furthermore, cell cultures inside the chip have sustained and controlled biochemical microenvironment due to easy control of conditions like pH, oxygen, nutrition, and temperature. Finally, real-time monitoring of cell can be achieved by introducing miniature electrodes (Kiilerich-Pedersen and Rozlosnik 2012).

Impedance spectroscopy can be used to check the toxicity of viruses (McCoy and Wang 2005). Drugs (Arndt et al. 2004; Hondroulis et al. 2010; Solly et al. 2004) and heavy metals (Chaubey and Malhotra 2002; Liu et al. 2007) effects can also be studied. Fig. 6.9b displays an eight gold electrode array used for nanotoxicity assays using electrical impedance sensing. Indeed, the effect of these compounds leads to a change of morphology of the cells that in turn is transduced to a change of resistance as described by Kiilerich and co-workers in 2011 (Kiilerich-Pedersen et al. 2011). They used an all-polymer electrochemical biosensor constituted by conductive polymer PEDOT:TsO microelectrodes and electrochemical impedance spectroscopy as detection technique leading to a label-free method. They could measure the real-time response of the cell to cytomegalovirus and detect an infection within 3 h.

In 2012, Larsen and Taboryski (2012) have described an electrochemical chip for the detection of transmitter release from cells. PEDOT:tosylate was used as electrode material and amperometry as detection technique. Cells were cultured inside the channel and they studied the response to potassium of these cells which release transmitters. In this study, real-time measurements could be obtained.

Such kind of sensors can also be employed to study circulating cancer cells as described by Arya et al. in 2012 (Arya et al. 2012). Gold electrodes were modified with anti-EpCAM (25 μm-diameter) in order to capture MCF7 cells. The capture event, monitored by EIS, led to a change of around $2.2 \times 10^7\,\Omega$ in the impedance value.

In 2012, Abdolahad et al. (2012) reported the use of carbon nanotubes (CNTs) to catch cancer cells. These CNTs were vertically aligned and used also as electrode material. As cancer cells have different mechanical properties, they were selectively entrapped by the CNTs and impedance changes occurred within 30 s. The device could detect the cancer cells with a concentration as low as 4,000 cells/cm^2 on its surface and with a sensitivity of $1.7 \times 10^{-3}\,\Omega\ \mathrm{cm}^2$.

6.4 Conclusions

In this chapter, we have reported the different fabrication methods to produce an electrochemical biosensor which can be integrated into microfluidics platforms as well as the classical techniques of detection that can be used. The use of microfluidics allows to such sensors reducing cost, sample amount, and assay time.

The quality of the electrodes and their integration inside a microfluidic platform are key factors in the performance of the biosensing system. While sputtering of metals ensures high quality thin film electrodes with an efficient sealing screen printing and ink-jet printing offer low cost electrode integration capabilities. Nevertheless, thick film electrodes produced by screen-printing technology have drawback related to sealing (leakages may occur) while ink-jet printing allows for thinner electrodes only for few commercially available inks that can fulfill strict particle size requisites.

Electrochemical biosensors integrated into microfluidics combine advantages of electrochemical biosensors with the versatility of sample introduction, possible separations, incubation steps, and detection in the same platform. Due to such versatility these systems are being used in various applications that include enzymatic biosensors, DNA sensing, immunosensing, and cells detection/studies.

Microfluidic-based electrochemical biosensors are simple in terms of easy integration of electrodes. They can offer highly sensitive biosensing systems due to the use of electrocatalytic materials that range from enzymes to nanomaterials (nanoparticles, quantum dots, carbon nanotubes, grapheme, etc.) coupled to various techniques including voltammetry, impedance spectroscopy (for label-free detection), etc. Such advantages are making these devices very interesting for a variety of applications with interest in diagnostic, safety and security beside other applications.

The use of microfluidic-based electrochemical biosensors in real samples analysis is strongly related to solving electrode fouling problems during their multiple uses or continuous monitoring applications. In addition, full integration of biosensing system parts (including detector/reader and related electronics) so as to achieve low cost and easy to use point of care devices for in-field applications requires additional efforts in multidisciplinary areas.

Acknowledgment This work was supported by MICINN project MAT2011-25870 and E.U. through FP7 "NADINE" project (contract number 246513).

References

Abdolahad M, Taghinejad M, Taghinejad H, Janmaleki M, Mohajerzadeh S (2012) A vertically aligned carbon nanotube-based impedance sensing biosensor for rapid and high sensitive detection of cancer cells. Lab Chip 12(6):1183–1190

Arndt S, Seebach J, Psathaki K, Galla HJ, Wegener J (2004) Bioelectrical impedance assay to monitor changes in cell shape during apoptosis. Biosens Bioelectron 19(6):583–594

Arya SK, Lee KC, Bin Dah'alan D, Daniel RARA (2012) Breast tumor cell detection at single cell resolution using an electrochemical impedance technique. Lab Chip 12(13):2362–2368

Azevedo AM, Martins VC, Prazeres DM, Vojinović V, Cabral JM, Fonseca LP (2003) Horseradish peroxidase: a valuable tool in biotechnology. Biotechnol Annu Rev 9:199–247

Bowden N, Brittain S, Evans AG, Hutchinson JW, Whitesides GM (1998) Spontaneous formation of ordered structures in thin films of metals supported on an elastomeric polymer. Nature 393 (6681):146–149

Brena B, Gonzalez-Pombo P, Batista-Viera F (2013) Immobilization of enzymes: a literature survey. Methods Mol Biol 1051:15–31. doi:10.1007/978-1-62703-550-7_2

Chaubey A, Malhotra BD (2002) Mediated biosensors. Biosens Bioelectron 17(6–7):441–456

Chen IJ, White IM (2011) High-sensitivity electrochemical enzyme-linked assay on a microfluidic interdigitated microelectrode. Biosens Bioelectron 26(11):4375–4381

Crosby JN, Hanley RS (1977) Chemical vapor deposition, US Patent 4250210A

Fang AP, Ng HT, Li SFY (2003) A high-performance glucose biosensor based on monomolecular layer of glucose oxidase covalently immobilised on indium-tin oxide surface. Biosens Bioelectron 19(1):43–49

Ferguson BS, Buchsbaum SF, Wu TT, Hsieh K, Xiao Y, Sun R, Soh HT (2011) Genetic analysis of H1N1 influenza virus from throat swab samples in a microfluidic system for point-of-care diagnostics. J Am Chem Soc 133(23):9129–9135

Ferrigno R, Lee JN, Jiang XY, Whitesides GM (2004) Potentiometric titrations in a poly(dimethylsiloxane)-based microfluidic device. Anal Chem 76(8):2273–2280

Fischer DJ, Hulvey MK, Regel AR, Lunte SM (2009) Amperometric detection in microchip electrophoresis devices: Effect of electrode material and alignment on analytical performance. Electrophoresis 30(19):3324–3333

Grieshaber D, MacKenzie R, Voros J, Reimhult E (2008) Electrochemical biosensors—sensor principles and architectures. Sensors 8(3):1400–1458

Guan JG, Miao YQ, Zhang QJ (2004) Impedimetric biosensors. J Biosci Bioeng 97(4):219–226

Heller A (1996) Amperometric biosensors. Curr Opin Biotechnol 7(1):50–54

Hondroulis E, Liu C, Li CZ (2010) Whole cell based electrical impedance sensing approach for a rapid nanotoxicity assay. Nanotechnology 21(31)

Jang YH, Oh SY, Park JK (2006) In situ electrochemical enzyme immunoassay on a microchip with surface-functionalized poly(dimethylsiloxane) channel. Enzyme Microb Technol 39 (5):1122–1127

Jensen GC, Krause CE, Sotzing GA, Rusling JF (2011) Inkjet-printed gold nanoparticle electrochemical arrays on plastic. Application to immunodetection of a cancer biomarker protein. Phys Chem Chem Phys 13(11):4888–4894

Kell DB, Davey CL (1990) Conductimetric and impedimetric devices. In: Cass AEG (ed) Biosensors: a practical approach. IRL, Oxford, pp 125–154

Kiilerich-Pedersen K, Rozlosnik N (2012) Cell-based biosensors: electrical sensing in microfluidic devices. Diagnostics 2(4):83–96

Kiilerich-Pedersen K, Poulsen CR, Jain T, Rozlosnik N (2011) Polymer based biosensor for rapid electrochemical detection of virus infection of human cells. Biosens Bioelectron 28(1):386–392

Kim JH, Kang CJ, Kim YS (2005) Development of a microfabricated disposable microchip with a capillary electrophoresis and integrated three-electrode electrochemical detection. Biosens Bioelectron 20(11):2314–2317

Komuro N, Takaki S, Suzuki K, Citterio D (2013) Inkjet printed (bio)chemical sensing devices. Anal Bioanal Chem 405(17):5785–5805

Larsen ST, Taboryski R (2012) All polymer chip for amperometric studies of transmitter release from large groups of neuronal cells. Analyst 137(21):5057–5061

Lauks IR (1989) Reference electrode, WO Patent 1989007758A1

Li DP, Sutton D, Burgess A, Graham D, Calvert PD (2009) Conductive copper and nickel lines via reactive inkjet printing. J Mater Chem 19(22):3719–3724

Li JT, Ye F, Vaziri S, Muhammed M, Lemme MC, Ostling M (2013) Efficient inkjet printing of graphene. Adv Mater 25(29):3985–3992

Lin FYH, Sabri M, Alirezaie J, Li DQ, Sherman PM (2005) Development of a nanoparticle-labeled microfluidic immunoassay for detection of pathogenic microorganisms. Clin Diagn Lab Immunol 12(3):418–425

Lin P, Luo XT, Hsing IM, Yan F (2011) Organic electrochemical transistors integrated in flexible microfluidic systems and used for label-free DNA sensing. Adv Mater 23(35):4035–4040

Liu DJ, Perdue RK, Sun L, Crooks RM (2004) Immobilization of DNA onto poly(dimethylsiloxane) surfaces and application to a microelectrochemical enzyme-amplified DNA hybridization assay. Langmuir 20(14):5905–5910

Liu QJ, Cai H, Xu Y, Xiao LD, Yang M, Wang P (2007) Detection of heavy metal toxicity using cardiac cell-based biosensor. Biosens Bioelectron 22(12):3224–3229

Lowe CR (1984) Biosensors. Trends Biotechnol 2(3):59–65

Maattanen A, Vanamo U, Ihalainen P, Pulkkinen P, Tenhu H, Bobacka J, Peltonen J (2013) A low-cost paper-based inkjet-printed platform for electrochemical analyses. Sens Actuators B Chem 177:153–162

Malaquin L, Limoges B, Goulpeau J, Marchal D, Kivlehan F, Le Nel A, Mavre F, Viovy JL, Taniga V, Miserere S, Mottet G (2010) Real time electrochemical DNA quantification in a COC lab on a chip: towards low-cost diagnosis of nosocomial infections. In: 14th International conference on miniaturized systems for chemistry and life sciences 2010, MicroTAS 2010 3:2053-2055

Marin S, Merkoci A (2009) Direct electrochemical stripping detection of cystic-fibrosis-related DNA linked through cadmium sulfide quantum dots. Nanotechnology 20(5)

Marrakchi M, Dzyadevych SV, Namour P, Martelet C, Jaffrezic-Renaulta N (2005) A novel proteinase K biosensor based on interdigitated conductometric electrodes for proteins determination in rivers and sewers water. Sens Actuators B Chem 111:390–395

Mayorga-Martinez CC, Hlavata L, Miserere S, Lopez-Marzo A, Labuda J, Pons J, Merkoci A (2014) An integrated phenol 'sensoremoval' microfluidic nanostructured platform. Biosens Bioelectron 55:355–359

McCoy MH, Wang E (2005) Use of electric cell-substrate impedance sensing as a tool for quantifying cytopathic effect in influenza A virus infected MDCK cells in real-time. J Virol Methods 130(1–2):157–161

Medina-Sanchez M, Miserere S, Morales-Narvaez E, Merkoci A (2014) On-chip magneto-immunoassay for Alzheimer's biomarker electrochemical detection by using quantum dots as labels. Biosens Bioelectron 54:279–284

Mehrvar M, Abdi M (2004) Recent developments, characteristics, and potential applications of electrochemical biosensors. Anal Sci 20(8):1113–1126

Moussy F, Harrison DJ (1994) Prevention of the rapid degradation of subcutaneously implanted Ag/AgCl reference electrodes using polymer-coatings. Anal Chem 66(5):674–679

Peckova K, Musilova J, Barek J (2009) Boron-doped diamond film electrodes-new tool for voltammetric determination of organic substances. Crit Rev Anal Chem 39(3):148–172
Perelaer J, de Laat AWM, Hendriks CE, Schubert US (2008) Inkjet-printed silver tracks: low temperature curing and thermal stability investigation. J Mater Chem 18(27):3209–3215
Puri N, Sharma V, Tanwar VK, Singh N, Biradar AM, Rajesh (2013) Enzyme-modified indium tin oxide microelectrode array-based electrochemical uric acid biosensor. Prog Biomater 2(1):1–7
Reichelt K, Jiang X (1990) THE preparation of thin-films by physical vapor-deposition methods. Thin Solid Films 191(1):91–126
Rius-Ruiz FX, Bejarano-Nosas D, Blondeau P, Riu J, Rius FX (2011) Disposable planar reference electrode based on carbon nanotubes and polyacrylate membrane. Anal Chem 83(14):5783–5788
Sheppard NF, Tucker RC, Wu C (1993) Electrical-conductivity measurements using microfabricated interdigitated electrodes. Anal Chem 65(9):1199–1202
Soldatkin OO, Kucherenko IS, Pyeshkova VM, Kukla AL, Jaffrezic-Renault N, El'skaya AV, Dzyadevych SV, Soldatkin AP (2012) Novel conductometric biosensor based on three-enzyme system for selective determination of heavy metal ions. Bioelectrochemistry 83:25–30
Solly K, Wang XB, Xu X, Strulovici B, Zheng W (2004) Application of real-time cell electronic sensing (RT-CES) technology to cell-based assays. Assay Drug Dev Technol 2(4):363–372
Swensen JS, Xiao Y, Ferguson BS, Lubin AA, Lai RY, Heeger AJ, Plaxco KW, Soh HT (2009) Continuous, real-time monitoring of cocaine in undiluted blood serum via a microfluidic, electrochemical aptamer-based sensor. J Am Chem Soc 131(12):4262–4266
Tortorich R, Choi J-W (2013) Inkjet printing of carbon nanotubes. Nanomaterials 3(3):453–468
Xiong ZT, Liu CQ (2012) Optimization of inkjet printed PEDOT:PSS thin films through annealing processes. Org Electron 13(9):1532–1540

Chapter 7
Applications of Paper-Based Diagnostics

Muhammad Safwan Akram, Ronan Daly, Fernando da Cruz Vasconcellos, Ali Kemal Yetisen, Ian Hutchings, and Elizabeth A.H. Hall

Abstract Paper has been used for applications in analytical and bioanalytical devices for more than a century, owing to its low cost due to its ubiquitous nature. Paper, a cellulosic material, presents several attractive attributes that render it useful in the construction of devices including: biodegradability, biocompatibility, worldwide abundance, chemical stability, three-dimensional fibrous structure, inertness to commonly-used reagents, ease of production and modification. Due to these characteristics, paper is one of the most widely researched substrates for the construction of low-cost disposable devices and sensing platforms. This chapter reviews the changing economic landscape, including the demand for low-cost diagnostics and a price comparison with other inexpensive substrates. The properties of paper, manufacturing challenges, labelling chemistries are also discussed along with the historical trends in marketed, paper-based devices.

7.1 Changing Trends in the Healthcare and Pharmaceutical Sector

It is not hard to predict that "diagnostics" will take the centre stage for driving innovation forward in the world of healthcare. This is noted in recent trends reported by MedTech Europe, as shown in Fig. 7.1a. Traditionally, paper-based diagnostic platforms have been pitched as a saviour for developing nations, providing access to inexpensive point-of-care diagnostics. However the truth is that they are also extremely relevant to the developed economies as well. One major trend that is shaping the healthcare market is that people are living longer and hence more money is spent on their healthcare as they grow older (Fig. 7.1b) (Hagist and

M.S. Akram (✉) • F. da Cruz Vasconcellos • A.K. Yetisen • E.A.H. Hall
Department of Chemical Engineering & Biotechnology, Institute of Biotechnology, University of Cambridge, Tennis Court Road, Cambridge CB2 1QT, UK
e-mail: safwan@cantab.net

R. Daly • I. Hutchings
Department of Engineering, Institute of manufacturing, University of Cambridge, 17 Charles Babbage Road, Cambridge CB3 0FS, UK

J. Castillo-León, W.E. Svendsen (eds.), *Lab-on-a-Chip Devices and Micro-Total Analysis Systems*, DOI 10.1007/978-3-319-08687-3_7

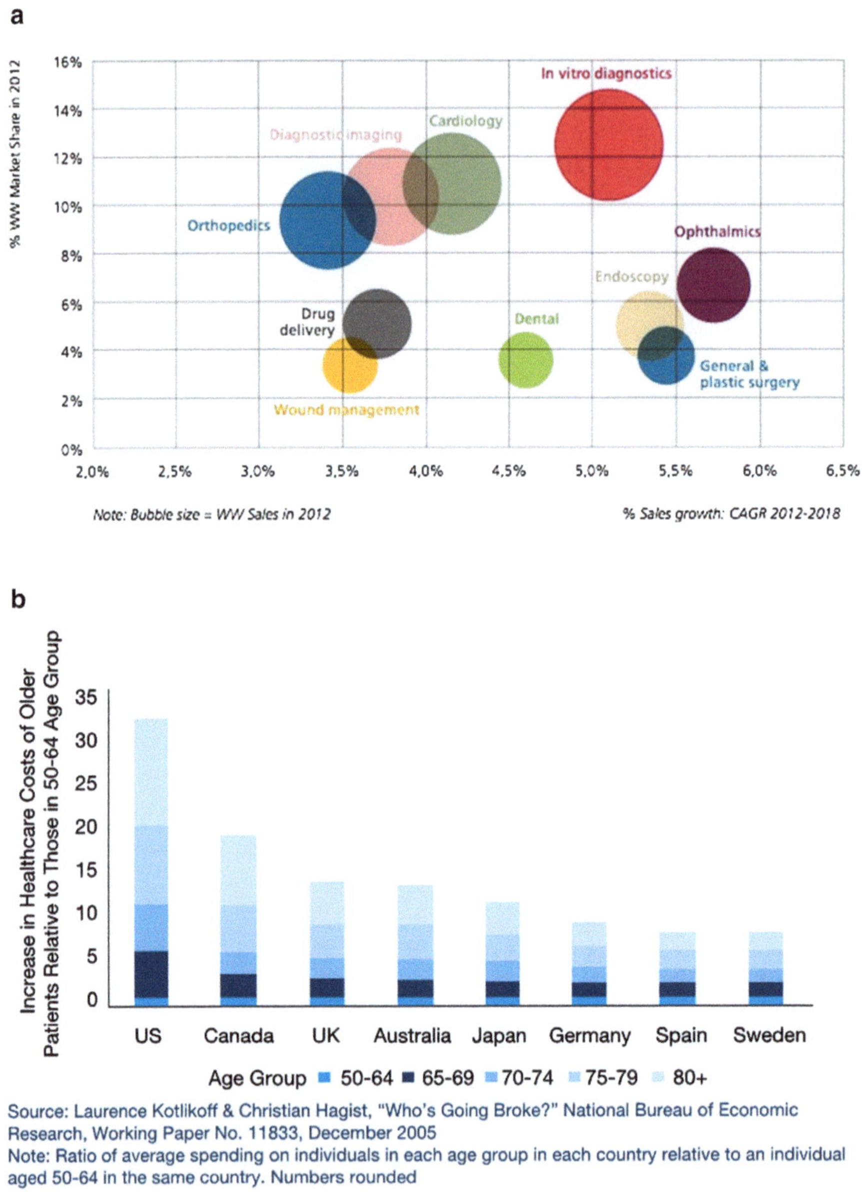

Fig. 7.1 (**a**) Medical Technology market estimates by area and sales growth, *x*-axis is in percentage of growth in sales while *y*-axis is the increase in percentage of market share (*Source*: MedTech Europe). (**b**) Consumption of healthcare budget by older people is considerably higher. (*Source*: PWC 2007 Pharma 2020 The Vision)

Kotlikoff 2005; PriceWaterHouseCoopers 2007). By 2020 about 720 million people (9.4 % of the world's population) will be aged 65 or older. This is true in both the developed and the developing world. Over the age of 75, 4 in 5 people currently take at least one prescribed medicine, with 36 % taking four or more (Assembly 2005). The population living with neurodegenerative diseases is on the rise, e.g. Alzheimer's disease afflicts 5.3 million people alone in the USA (costing $172 million per year) and this number is estimated to quadruple by 2050 (Finkbeiner 2010; Thies and Bleiler 2011). Several companies working to develop drugs are finding it very hard to recover the high costs associated with the treatments. For example the UK National Institute of Clinical Excellence (NICE) has ruled that Aricept (Eisai, Pfizer), Exelon (Novartis) and Reminyl (Shire) should not be used as the first line of defence because "these drugs didn't make enough of a difference" to justify their cost, restricting their use only for moderate to severe symptoms. Such restrictions provide drivers for the investigation of biomarkers and put a strong case for companion diagnostics for the drugs in clinical trials, which can be used to predict improvement to the health of patients taking the drug and monitor the side effects. Ignoring this can dramatically effect the reimbursement of the drug due to the subsequent lack of appropriate market access. With the tremendous increase in healthcare budgets around the globe, payers (Governments and Health Insurers) are asking to pay for performance and are demanding demonstrable benefits from the treatment they support. For the first time, pharmaceutical companies have to prove that their products are effective in alleviating symptoms beyond the confines of controlled clinical trials environments. Such demands have given a new life to the field of theranostics or companion diagnostics. However, for drug companies, performance is a measure borne out of financial and regulatory pressure, it means that pharmaceutical companies do not have a huge leeway and would have to adjust the cost of developing diagnostics out of their operating costs. This means that low-cost diagnostics are required as much in the developed world as in developing countries. It should be kept in mind that any companion diagnostic kit, for the purpose of the clinical trial or for the continuing phase four trials, would be limited by the number of patients using that product. This means that the traditional concept of economies of scale does not apply. Instead low-cost alternatives are required which are configurable at the point of manufacture. BiognostiX, one such configurable platform, is described in this chapter and was developed by a consortium of the University of Cambridge, UK with four companies: FFEI (UK), Parco Tecnologico Padano (Italy), Prionics (Switzerland) and Proteomika (Spain).

7.1.1 Low-Cost Diagnostics in the Developing World

The importance of point-of-care diagnostics in the developing world cannot be overemphasised. Inhabitants lack access to basic health care infrastructure and centralised laboratory facilities, and suffer from a shortage of trained healthcare personnel (Varmus et al. 2003). Low-cost point-of-care diagnostic devices have the potential to reach these under-served communities at the bottom of the pyramid (Hu et al. 2014). Their potential value is not limited only to healthcare; devices are being developed to determine quality of food at the point-of-sale or use, for monitoring of livestock and for environmental safety. The most plausible route to alleviate poverty from these regions is to offer them trade opportunities with the developed world. However, this trade cannot involve compromise in product quality, especially in the area of food products. Low-cost diagnostics would allow farmers in the developing world to validate their products and compete better in the world market. As an example, in the food market of the developing world an enterohaemorrhagic strain of *E. coli* (O157:H7) is of major concern; it is transmitted via the faecal–oral route through undercooked ground meat (beef and pork). Similar concerns are shared for various subspecies of salmonella in undercooked poultry meat and eggs (Organization 2002).

One major concern is the quality of drinking water, the source of a significant proportion of the prevalent diseases in these communities, to the extent that a child dies every 21 s from waterborne disease (water.org 2014). Resurgence of cholera after the earthquake in Haiti in 2010 highlighted the fact that if developing communities are not empowered to take care of their own health then nobody is safe, and if the situation continues on the same path we may start to see epidemics being converted into pandemics (Cravioto et al. 2011). This situation requires populations in the developing world to be equipped with low-cost methods of detection and diagnosis. Market demands are clear and the challenge presented to low-cost diagnostics is to be rapid enough (i.e. with low turnover time) to aid in screening of a large population for multiple pathogens, and at the same time quantitative enough to help clinicians with diagnosis. Echoing these challenges, the World Health Organization (WHO) has issued a criterion for diagnostic devices in the developing world with the acronym ASSURED: Affordable, Sensitive, Specific, User-friendly, Rapid and robust, Equipment free and Deliverable to end-users (Peeling et al. 2006).

7.1.2 Economic Rationale for the Use of Paper as a Substrate

For last 5 years or so, paper has been the poster child for researchers working to develop low-cost diagnostics. It is not hard to find the reasons as paper's potential to be an excellent substrate for microfluidic devices is enormous owing to its ubiquitous nature, low cost, compatibility with a myriad of chemicals and ability to transport liquids using just the capillary forces without any external assistance. We recently performed a cost-based comparison between paper substrates and silk which is another low cost alternative (Yetisen et al. 2013). The most expensive part of the manufacturing process is fabrication of the microfluidic channels and even then, paper-patterning costs can be as low as <\$0.01 per device (Martinez et al. 2009). This is considerably lower than the cost of microfluidic patterns made from moulded plastic (Personal Communication with Dendukuri

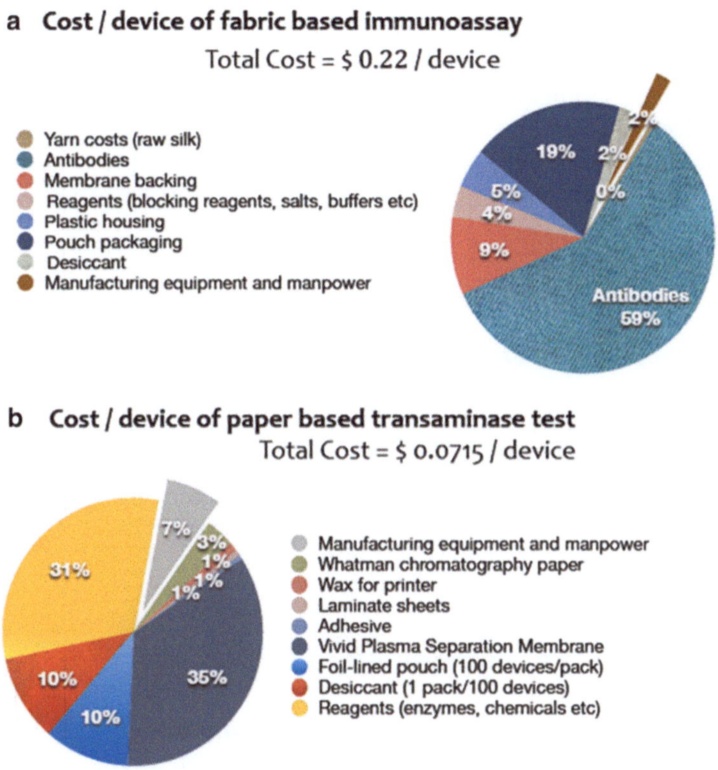

Fig. 7.2 Cost analysis of fabric and paper-based devices. (**a**) The breakdown cost/device is given for Fabric based chips developed by Achira labs (India). The total cost of a device is $0.22. The costs are amortised over ten million devices. (**b**) The breakdown of Cost/device of paper-based transaminase test is given. The total costs per tests are calculated to be $0.0715. These costs are based upon Indian market and in US it would two to threefold higher. The figure is drawn from a supplementary table give by Diagnostics for All (Pollock et al. 2012). [Adapted from Yetisen et al. (2013)]

28 December 2012, CEO of Achira Labs. Interview on feasibility of Fab-chips) and is also inclusive of the price of the paper. Figure 7.2a shows the breakdown of costs involved in a fabric-based immunoassay compared with cost per device for a paper-based transaminase test (Fig. 7.2b). If costs are amortised over ten million tests, then paper offers savings of $1.48 million; however, one should not ignore the importance of the costs of other materials e.g. the most expensive part in the paper-based aminase test is the plasma separation membrane, while the most expensive part of the fabric-based immunoassay (Bhandari et al. 2011) is the antibodies itself as they are much more expensive than enzymes.

7.2 Properties of Paper-Based Devices

The growing interest in and awareness of paper-based devices may be attributed to the many desirable properties of paper, some of which are presented in Table 7.1.

The main properties of paper as a device substrate compared to traditional materials (glass (silicon-dioxide-based) and plastic (polymer-based)) are presented in Table 7.2.

7.2.1 Paper-Based Device Assembly

Production techniques for paper-based devices usually involve some form of cutting, folding, writing, printing, plotting, dip-coating, spray-coating, and sputtering or lithography. Examples of some of the most commonly used techniques to prepare paper-based devices are presented in Table 7.3.

7.3 Current Status of Paper-Based Devices

7.3.1 The Journey of Paper-Based Microfluidics

The origins of paper-based microfluidics date back to the early 1900s, where impregnation of paper with hydrophobic materials was used to minimise the diffusion of reagents and to prevent cross-contamination of reaction zones in the paper matrix (Dieterich 1902; Yagoda 1937, 1938). The first application found its way to the market in the form of qualitative spot testing where paraffin wax was used to confine the sample to a small area (Fig. 7.3a, c) (Yagoda 1937). Paraffin wax became the material of choice owing to its hydrophobicity and inertness to commonly used chemical reagents, its ease of printing to create diverse hydrophobic patterns and channels in its molten form. Further help came from its easier two step manufacturing process where embossing tools and moulds are warmed and dipped into paraffin before being stamped onto the filter paper. Spread of the paraffin wax, the temperature of the dye, the pressure and the time of contact are optimised for the desired paper type and its pore size distribution. Early applications involved devices that could detect low concentrations of nickel and copper salts, with speculations that similar platforms might be useful for pH, environmental testing and biological analyses of urine and blood (Johnson 1967). In late 1940s, the attractive attributes of building such devices were realised: faster analysis and reducing the reagent consumption by using smaller quantities (Müller and Clegg 1949). Around the same time, multiplex immunoassays were developed capable of running multiple tests in parallel (Fig. 7.3b). Although these studies gained limited attention from

Table 7.1 Properties of paper-based devices[a]

	Property	Importance for devices
Natural organic material	Paper is primarily constituted of cellulose, which is one of the most abundant natural organic material in the world	Readily available, renewable resource, low to medium cost
	Biodegradable	Rapidly degradable by microorganisms (Couderc et al. 2009)
	Reusable	Recyclable components (Noh and Phillips 2010a)
	Disposable	Easily disposed of by incineration, reducing/eliminating issues relating to contamination with biological and chemical samples (Martinez 2011)
Physical properties	Flexible	Non-linearly positioned or folded paper devices will maintain their structural integrity
	Thickness (tens to hundreds of micrometres)	Low (microliter) volumes of fabrication material and small sample aliquots are required (Tseng et al. 2012)
	Lightness (~10 mg/cm^2)	Device portability (Jagadeesan et al. 2012)
	High specific stiffness	Device robustness (Werner et al. 2010)
	Thermally stable	Resistance to mild temperature variations
Chemical properties	Chemically and biologically inert	Inertness (Kouisni and Rochefort 2009) is advantageous in the immobilisation and maintaining integrity of antibodies and proteins (Kim et al. 2007)
	Easy to sterilise	Adequate for biomedical and other sterile applications (Su et al. 2008)
	Presents chirality	Advantageous in the immobilisation of antibodies and proteins (Kim et al. 2007)
Porous fibrous structure	High surface-to-volume ratio	Allows for the immobilisation of a large number of functional molecules, such as enzymes, DNA and molecular probes. Also provides large surface areas for transducers
	Capillary action	Provides the capability of wicking fluid samples and permits fluid flow in all directions (Hossain et al. 2009)
	Absorbency	Facilitates storage, delivery, mixing steps to allow the desirable flow of samples through channels and reaction zones (Fu et al. 2012; Bracher et al. 2009; Gu et al. 2011)
	Liquid and gas permeability	Allows the permeability of liquids and diffusion of gases through the devices, increasing the operating area of the device (Xu et al. 2011)
	3-D network structure	Permits the implementation of separation, filtration and mixing in the device architecture (Martinez et al. 2008a)

[a]Adapted from Nery and Kubota (2013)

Table 7.2 Paper-based substrate in comparison to glass, silicon and PDMS-based substrates[a]

	Substrate type		
Characteristics	Paper	Glass (silicon-dioxide-based materials)	Plastic e.g. polydimethyl (siloxane) (PDMS)
Availability	High	High	High
Physical stability	Moderate to high	High	High
Chemical stability	Moderate	High	Moderate to high
Volume structure	Porous	Solid	Solid
Ease of surface functionalization	Low to moderate	Moderate	Moderate
Surface-to-volume ratio	High	Low	Low
Biodegradability	Yes	Yes	Yes
Disposability	Yes	No	No
Biocompatibility	Yes	Yes	Yes
Mass-producibility	Yes	Yes	Yes
Homogeneity of material	Low	High	Moderate to high
Batch-to-batch consistency	Moderate	High	High
Sensitivity to moisture	High	No	No
Sensitivity to heat	Moderate	Low	Moderate
Fluid flow	Capillary	Capillary to forced	Capillary to forced
Surface roughness	High	Low	Moderate
Spatial resolution	Low to moderate	High	High
Cost	Low to moderate	High	Moderate to high
Initial capital investment required	Low to moderate	Moderate	Moderate to high

[a]Adapted from Nery and Kubota (2013)

industrial partners, they showed proof-of-concept for early paper-based microfluidics.

Research within the in vitro diagnostics industry focused on preventing cross-contamination in dipstick tests made out of polymers like polycarbonate, poly (methylmethacrylate) (PMMA) and cyclic olefin copolymer (COC). Forming hydrophobic barriers between reaction zones, strategies to absorb run over liquid using bibulous layers (1979), and physical separation of adjacent reagent zones were a few solutions. Materials that are impervious to water were widely used: such as ethyl cellulose, silicones, polystyrene, rosin, waxes, paraffin, printer varnish and cellulose esters. These developments led to the introduction of, rapid diagnostic assays in consumer markets. The first commercial test was a paper-based diabetes dipstick test, which allowed semi-quantitative analysis of glucose in urine

Table 7.3 Commonly used assembly techniques for the preparation of paper-based devices[a]

Techniques		Advantages	Disadvantages
Cutting	Manual (Wang et al. 2010) Automatic Laser (Bracher et al. 2009)	Simple and versatile; does not expose paper to contaminants	Device features generally display low resolution features, with the exception of laser cut devices
Folding	Origami (Liu et al. 2012; Liu and Crooks 2011)	3D devices are attainable to facilitate multiplexing	Edges may result in undesirable flow is not prepared adequately
Taping	Taping/gluing (Martinez et al. 2008b)	Assembly and placement location are maintained for 3D paper devices	Chemicals in the tape/glue may interfere with the paper device mechanism
Writing	Manual	Ease of fabrication, low resources required	Low resolution
Printing	Ink-jet printing (Abe et al. 2008; Balu et al. 2009; Li et al. 2012a)	Fast, automated, computer assisted design and production	Nozzle clogging, temperature dependence of dispensing reagents, costly printers
	Wax printing (Carrilho et al. 2009a; Ge et al. 2012)	Hydrophobic barriers may be readily fabricated	Wax printers are not widely used
	Screen-printing (Dungchai et al. 2011)	Device channels may be constructed with varying thicknesses	Prototyping is more difficult, and this process requires a mesh
Plotting	Auto-plotting (Bruzewicz et al. 2008b; Araujo et al. 2012)	Displays higher flexibility in terms of depositing material on various surfaces, compared to traditional printing methods	Control over dispensing material may be difficult to control, plotters are generally costly
Dip-coating	Immersion (Songjaroen et al. 2012) LbL deposition (Gao et al. 2013; Alkasir et al. 2012)	Facile technique, conformal and uniform coatings achieved, widely used and low-cost	Producing times may be lengthy, hindering mass producible capabilities
Spray-coating	Atomisation LbL spraying	Facile technique, faster compared to traditional dip-coating	Custom equipment is generally required
Sputtering	Lankelma et al. (2012)	Versatile technique to dispense metallic materials on the device's surface	Masks and patterns may be required to produce the desired sputtered patterns
Lithography	Photolithography (Nie et al. 2010)	Facile and versatile method to produce devices features	Requires masks and chemical development processes, which may leave contaminants
Heating	i.e. for wax channel formation (Carrilho et al. 2009a)	Wax channels are extended throughout the paper thickness, e.g. producing a barrier to hydrophilic components in aqueous mixtures	Wax is not uniformly spread

(continued)

Table 7.3 (continued)

Techniques		Advantages	Disadvantages
UV exposure	He et al. (2013)	Sterilisation is achieved; component crosslinking is possible without harsh chemistry	Safety precautions need to be in place to avoid undesirable UV exposure
Plasma treatment	O_2 plasma (Martinez et al. 2008b)	Assists with surface functionalisation; assists in sterilising the device	Traditional plasma treatment equipment are small, operate in batches, and require oxygen specific vacuum pumps to operate properly

[a]Adapted from Nery and Kubota (2013) and Yetisen et al. (2013)

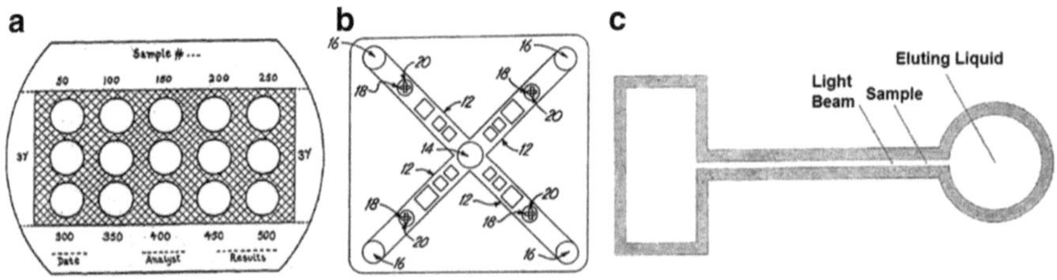

Fig. 7.3 Earlier paper-based devices. (**a**) A paper-based microplate assay. Adapted with permission from ref. Yagoda et al. Copyright (1937) American Chemical Society. (**b**) A paper-based multiplex microfluidic device for simultaneously performing multiple competitive immunoassays from ref. US Patent 5707818 (1998). (**c**) A microfluidic sample inlet, channel and detection zone fabricated by impregnating paper with wax. Adapted with permission from ref. Muller and Clegg Copyright (1949) American Chemical Society

(Free et al. 1957). Cellulose paper remained components of lateral flow assays in the form sample and absorption pads but went out of favour for reactive membrane region in comparison to nitrocellulose. Nitrocellulose membranes became the material of choice due to their uniform pore size, hydrophilic nature, high protein binding capacity and ease of use in large scale manufacturing systems. The first point-of-care tests containing nitrocellulose were latex agglutination assays and radioimmunoassays (Plotz and Singer 1956; Berson and Yalow 1959). From 1970 to 1990, a number of molecular detection assays were developed (Southern 1975; Towbin et al. 1979; Goldberg 1980) including serological lateral flow tests (Hawkes et al. 1982). Such tests found applications in testing pregnancy based on human chorionic gonadotropin (hCG) (Vaitukaitis et al. 1972; Martinez et al. 2008b). Since then dipstick and lateral flow tests found wider applications in urinalysis, immunoassays, veterinary screening, food safety, environmental monitoring, bio-threat detection and drug abuse screening (1982; 1983; 1984; 1987; 1989; 1990; 1991; 1997). Today such dipstick tests are widely used for screening urinary tract infections (UTI), diabetes and kidney disorders) (Vaitukaitis et al. 1972; Hawkes et al. 1982; Ordway et al. 1997; 1982; 1983; 1984; 1987; Yazawa, Nukina et al. 1995; Chen, Stott et al. 1999).

These developments in use of paper as a substrate were concomitant with the myriad advances in ink jet printing and handling of small volumes through gas phase chromatography (GPC), high-pressure liquid chromatography (HPLC) and capillary electrophoresis (CE). Microfluidics also started to emerge in niche applications e.g. Zigmond developed a microfluidic device to visually study the chemotaxis of polymorphonuclear leucocytes as early as 1977 (Zigmond 1977). However, stimulus for the development of field deployable microfluidic systems was provided by handsome funding in the 1990s from the United States Department of Defense (USD) and Defense Advanced Research Projects Agency (DARPA) to develop detectors for biological and chemical threats.

Today there is a massive push in academia to design and fabricate sensors that are totally made out of paper. One such example is μPAD for the detection of glucose and proteins discussed in Sects. 7.5.1 and 7.5.2. Another very good example is the semi-quantitative low cost point-of-care test developed by Whitesides group to monitor liver function (Pollock et al. 2012). Figure 7.4 shows their assay that attempts to quantify level of aspartate aminotransferase (AST) and alanine aminotransferase (ALT) in blood in 15 min. This test is very important for AIDS patients with drug related hepatocytotoxicity and acts as a companion diagnostic for the drug "Nevirapine" against reverse transcriptase of HIV1. These tests are patterned in a circular disc shape on a square piece of paper in a 3D arrangement where top layer acts as a filter to remove red blood cells. Carrilho et al. (2009b) have developed microzone titer plates for immunoassays very similar to 96-well plastic plates commonly used for ELISA. The idea is to use the existing readers for final outcomes of the assays to minimise the shifting costs to inexpensive substrates.

7.4 Technical Achievements and Challenges

Biosensors and diagnostic systems are often compared using the key performance factors of selectivity, sensitivity range, accuracy, response time, recovery time and working lifetime (Eggins 2008). The challenge facing the field of paper-based diagnostic devices is to achieve a beneficial and sufficiently competitive combination level of these factors. However, this must also preclude compromising the advantages over alternative microfluidic or lab-based detection methods, namely using a low-cost, sustainable material with inherent fluid transport properties and an associated rich history of handling and process technologies. As highlighted in Sect. 7.1.3, such a combination presents the opportunity to provide a significant improvement in the cost and flexibility of point-of-care medical technologies, opening up advanced diagnostics to developing nations (Lei and Yang 2013; Li et al. 2012b; Martinez et al. 2009; Pelton 2009a). The small number of products implemented in such regions (and indeed globally) highlights the fact that serious

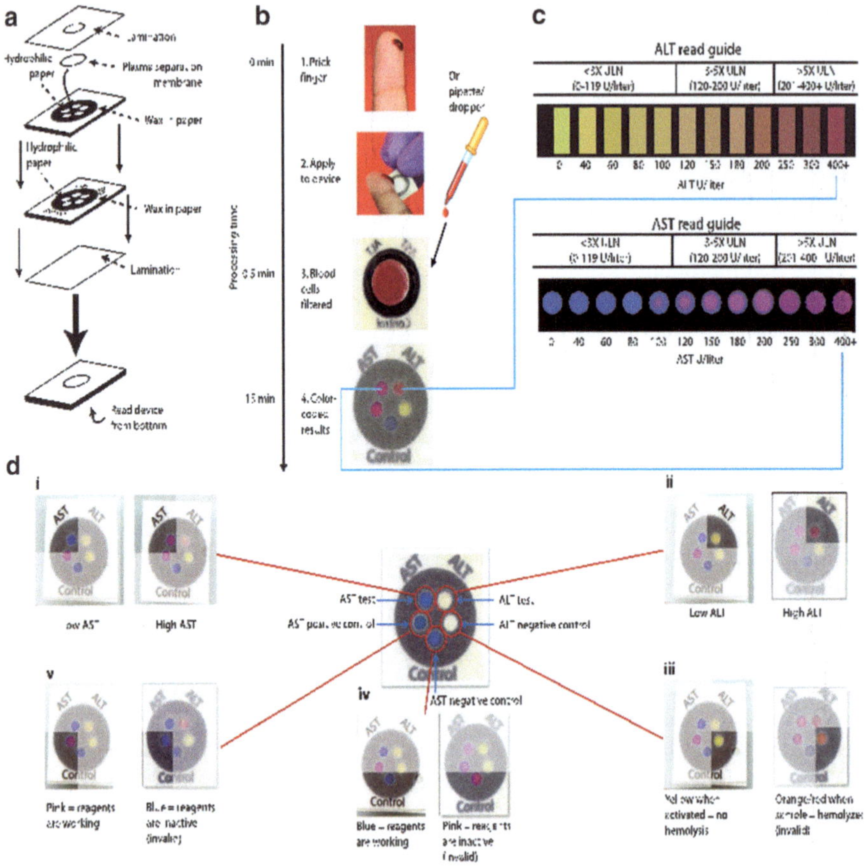

Fig. 7.4 Schematic of the paper-based AST/ALT test design and protocol. (**a**) The device consists of two layers of similarly patterned paper, a plasma separation membrane, and a laminated cover of polyester film. (**b, c**) A drop of whole blood (either a fingerstick specimen or 30 ml of a specimen obtained by venipuncture) is applied to the back of the device. Red and white blood cells are filtered out by the plasma separation membrane, whereas plasma wicks to the five detection zones through patterned hydrophobic channels in the paper (**b**). After 15 min, the AST and ALT test zones are matched to a colour read guide (**c**) to obtain a concentration value. Results are interpreted as being within one of three bins of values: <3× ULN (ULN defined as 40 U/l), 3–5× ULN, or >5× ULN. (**d**) Detailed schema of the paper-based transaminase test and possible colorimetric readouts for the five zones (*i–v*). A schematic of test and control zones (before receiving a sample) is shown in the centre of the figure. *i*) AST test zone: low/normal AST values (<80 U/l) result in a *dark blue colour*, whereas high AST values (>200 U/l) result in a *bright pink colour*. (*ii*) ALT test zone: low/normal ALT values (<60 U/l) result in a *yellow colour*, whereas high ALT values (>200 U/l) result in a deep *red colour*. (*iii*) ALT negative control zone: a change from *white* to *yellow* indicates appropriate device activation; in the event of sample hemolysis, the zone becomes *orange/red* and the device is read as invalid. (*iv*) AST negative control zone: the baseline *blue colour* remains unchanged if dye chemistry is functioning properly, whereas the zone becomes *bright pink* in the event of non-specific dye reaction and the device is read as invalid. (*v*) AST positive control zone: the zone changes from *blue* to *pink* if AST reagents are functioning properly but remains *dark blue* if either the reagents are not functioning or the zone is not activated, in which case the device is read as invalid. [Copied from Pollock et al. (2012) and copyrights belong to AAAS]

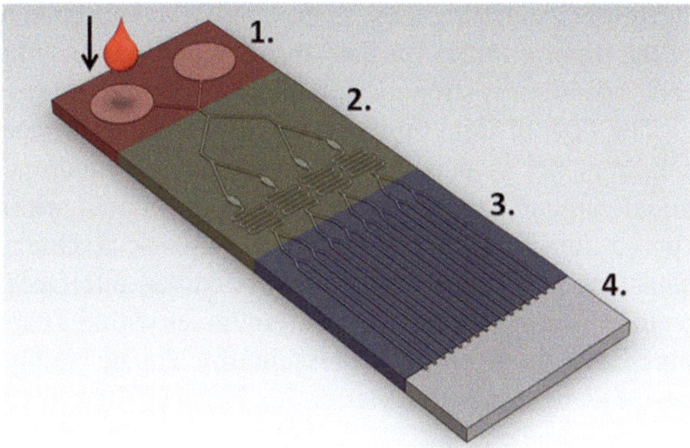

Fig. 7.5 Illustration of a paper-based diagnostic device with four key operations of (1) sample preparation and delivery to the fluidic channels, (2) microfluidic flow, additional separation and mixing steps, (3) reaction and capture of the analyte, and (4) sample collection to ensure that the full volume passes the reaction zone

underlying challenges remain at a number of stages in the value chain. It is evident from the large number of scientific publications on the subject and their broad scope that there are fundamental challenges at the basic scientific level from implementing specific test chemistries on to cellulose to ensuring a sufficiently robust fabrication technique.

The range and complexity of approaches taken to incorporate cellulose fibres into diagnostic devices varies considerably. Key review papers in the field of paper-based diagnostics most often describe the approaches in terms of the fabrication technique used to define the channel flow, as shown in Table 7.3 (Pelton 2009a; Li et al. 2012b; Yetisen et al. 2013). In this chapter we present a higher-level overview of paper-based diagnostics in which we explore the achievements and technological challenges in terms of four key operations common across the use of all diagnostic devices reviewed here:

Operation 1: Sample preparation, delivery to the coupon and subsequent delivery to the fluidic channels.

Operation 2: Transport of the sample through the device, including any required division or mixing of flows.

Operation 3: Capture and recognition of the analyte in the reaction zone with appropriate signal generation.

Operation 4: Collection of the entire sample by absorption for containment and to ensure that all of the analyte passes through and is tested in the reaction zones.

An illustration of these divisions is shown in Fig. 7.5 for a diagnostic device under development.[1] Each operation is examined in turn to identify the benefits achieved by using paper, the approaches reported in the literature and the key challenges that must be addressed to progress the technology to commercial use.

[1] www.biognostix.com.

Before examining these operations, it is important to consider a brief overview of the driving forces for flow through non-woven, fibrous cellulosic materials, as each operation is strongly influenced by the basic principles.

The flow of liquid through the pores between paper fibres can be treated in a simple model as analogous to flow through cylindrical capillary tubes. In a cylindrical capillary, the contact angle, θ, of a liquid within the small tube leads to a curved liquid–air interface, seen as a meniscus. Due to surface tension, there is a balancing pressure difference across curved interfaces, the pressure being greater on the concave side. Liquid flow continues within a capillary tube or between paper fibres until the hydrostatic pressure drop equates to this pressure difference. Equation (7.1) describes the relationship between the pressure drop, the capillary radius and the balance of surface tensions.

$$\Delta P = \frac{2\left(\gamma_{1/3} - \gamma_{1/2}\right)}{r} \tag{7.1}$$

where r is the radius of the cylindrical capillary, $\gamma\frac{1}{3}$ is the paper/air surface energy, $\gamma\frac{1}{2}$ is the paper/liquid interfacial energy, and ΔP is the pressure difference across the curved interface under static conditions. For a liquid to enter a capillary, a positive ΔP is required so the solid–liquid interfacial energy must be lower than that of the solid–air interface. A further development often used is that of the Washburn equation, which describes the rate of imbibition of aqueous solutions into a porous material, as shown by Eq. (7.2).

$$h_{\mathrm{p}}(t) = \sqrt{\frac{\gamma_{2/3} l_r}{2\eta} . \sqrt{t}, l = \gamma_{1/3} - \frac{\gamma_{1/2}}{\gamma_{2/3}}} \tag{7.2}$$

where h_{p} is the depth of liquid penetration in time, t, r is the capillary radius, η is the viscosity of the penetrating liquid and $\gamma\frac{2}{3}$ is the liquid–air interfacial energy.

The basic understanding of surface-tension driven flow is contained within these equations. While more complex numerical approaches are often employed, these models still underpin many of the contemporary approaches (Mullins et al. 2007; Hyväluoma et al. 2006; Jaganathan et al. 2008). The two equations illustrate very well the competition between the drive to minimise surface energy and the resistance to flow due to viscous forces. These equations will be referred to throughout the following discussion and will aid in the understanding of the challenges facing all four operations described below.

7.4.1 Sample Preparation, Dosing and Downstream Delivery

The first operation consists of choosing a sample source, preparing it for use in the diagnostic device and delivering it to the microfluidic channels. While the initial consideration is of course to select source (blood, urine, saliva, etc.) with a high likelihood of containing the target analyte, this sample must be prepared with the specific purpose of ensuring flow through a porous non-woven matrix. It is important therefore to consider (1) the sample preparation required ensuring compatibility with paper-based diagnostics, (2) the necessary accuracy of the dosing device and (3) the technology required to hold the dosed sample in the device and to meter it effectively to the second operation, namely flow along the microfluidic flow channels.

Sample Preparation

We can see from Eqs. (7.1) and (7.2) that absorption and flow rate are sensitive to both the surface tension of the liquid sample and any coatings functionalizing the paper. The advantage of using paper for the channel materials is that its reported surface energy of 60.8 mN/m (dispersive component, 25.9 mN/m and polar component, 34.9 mN/m (Cunha and Gandini 2010); lends itself naturally to the transport of aqueous biological samples, with the polar component specifically responsible for hydrophilicity. This is not the case with most polymeric materials, which have low polar components of surface free energy, e.g. polystyrene (6.1 mN/m), polycarbonate (6.5 mN/m), PTFE (1.6 mN/m).[2] For any chosen biological sample to be compatible with a microfluidic device, it must not compromise this advantage and therefore must have properties within appropriate limits of viscosity and surface tension. Samples reported in the literature are often highly dilute aqueous solutions. This ensures that both the surface tension and viscosity stay close to the ideal combination for capillary flow and diffusion. Difficulties lie in using more viscous samples, such as whole blood.

There are further underlying scientific challenges that must be tackled to make progress towards a practical device. To ensure that non-specific binding of proteins is sufficiently low, surface treatments with Bovine Serum Albumin (BSA), casein, Tween 20 or Triton X-100 are commonly carried out (Crowther 2009). However, such coatings can have a significant effect on the surface energy of the surface, e.g. ~20° reduction is observed in the contact angle for water on PMMA coated with BSA compared with the untreated polymer (Białopiotrowicz and Jańczuk 2001). Detergents inclusion assists with a number of processes, such as cell lysis and protein precipitation. However, addition of any detergent changes the surface tension of the liquid–air interface dramatically, even at the typically low levels at ~0.05% w/w occasionally present in such systems. Surface tension can drop from close to 72–35 mN/m (Niño and Patino 1998). This has two potentially important effects. First, the flow rate will change significantly (Hodgson and Berg 1988) and second, if channels are defined by hydrophobic barriers as noted in Table 7.3, they do not necessarily guard against imbibitions from the modified solution, leading to

---

[2] http://www.surface-tension.de/solid-surface-energy.htm.

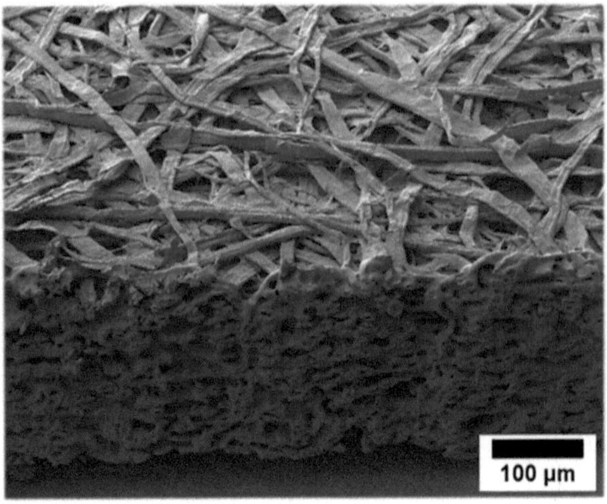

Fig. 7.6 Cross section of Whatman 1 CHR paper, prepared by laser-cutting, coated in Au-Pd and imaged using scanning electron microscopy. The scale bar is 100 μm

cross-contamination between channels or sample leakage. When designing a paper-based sensor, it is therefore essential to consider the batch-to-batch variations of these components in the samples and try to develop the platform robustness accordingly.

The rate of entry of a sample into a capillary is shown from Eq. (7.2) to be directly proportional to viscosity. However, untreated samples from humans and other animals exhibit a wide range of viscosity e.g. urine samples are most often close to the viscosity of water while saliva and blood are more viscous (and less consistent). Both the latter are highly non-Newtonian, with viscosities ranging from 2 to 4 cPs at high shear rate to a viscosity of the order of 100 cPs at zero shear rate (Stokes and Davies 2007; Brujan 2011). Blood is especially complex due to its heterogeneity: blood cells are sufficiently close in size to the pore dimensions that the sample needs to be treated as a fluid (plasma) containing discrete particles, rather than as a homogenous viscous fluid.

Finally, the porous matrix (as seen in Fig. 7.6) will often not allow the free flow of particulates above a certain size. While the majority of diagnostic systems rely on the flow of proteins, antibodies and materials soluble in the carrier fluid, traditional lateral flow assays often use suspended, conjugated coloured particles. When these are captured in the reaction zone, the increase in local concentration provides a visible reflected (or emission) output signal. In such cases, the preparation of the particle/analyte conjugate is a critical step and often occurs prior to dosing. In this case the sample is now a suspension of particles and must be carefully prepared to ensure that a highly stable, representative concentration is dosed into the device.

The particle size is also an essential parameter: e.g. conjugated latex beads of ~300 nm in diameter were unable to flow through the porous matrix in a cellulose-based lateral flow assays (Lappalainen et al. 2010), whereas the test was successful when employing conjugated nanoparticles. In a similar system, there is a significant interest in sensing the presence and concentration of bacteria for applications such as human health or food hygiene. The range of particle sizes and aspect ratios inherent in such samples is very broad and again must be considered if they are required to flow through a porous matrix. While some success has been demonstrated (Li et al. 2011) it may be important often to preprocess the sample and use only the cell lysate (Hossain et al. 2012) or include an additional stage within the device to pre-filter the sample (Yang et al. 2012; Songjaroen et al. 2012).

These considerations noted as part of the Operation 1 highlight the need to consider sample preparation as part of the design of the diagnostic device especially if attempting to develop a simple point-of-care testing arrangement for use in developing nations or field tests without the ease of access to preparation techniques.

Sample Delivery

The importance of the method by which the biological sample is applied to the point-of-care diagnostic device is often overlooked. However, it may lead to unintended effects or limitations. Dip-tests are used where sufficient volumes, sterile containers and appropriate waste-disposal are all available. These can take the form of single or parallel tests, as used in the commercial test for BSE from Prionics® (Prionics). Accurate, manual or auto-dispensing by a micropipette is also often used when quantitative results are needed. This ensures that a known sample volume passes through the reaction zone so that a quantitative interpretation can be carried out but this assumes some level of access to and training with such devices, which may not be feasible in some applications. Dosing can be directed to sample deposition zones on non-absorbent or absorbent materials. When delivered to a non-absorbent location (Tian et al. 2010), this leads to a standard microfluidic channel and then a paper region to carry out the reactions. This approach may allow complete transfer of the sample but also adds complexity to the manufacturing process. Paper is also discussed in the literature as a dosing pad and hence as a means of receiving the sample. The very rapid absorption ensures less time for contamination to occur through contact with either the user or the environment and also less time for any potential phase separation (e.g. surface active components migrate to the air interface). The ability to functionalize paper with dried materials prior to sample delivery, e.g. dried urine paper (Zava et al. 2013), also shows that label pick-up and conjugation can be achieved as part of the dosing technique, as opposed to pre-processing of the sample. By using existing understanding from the field of filtration, paper can also be designed or used in conjunction with the dosing step to remove components from the sample prior to moving to Operation 2 (Yang et al. 2012; Songjaroen et al. 2012). Metering of the sample to the second operation is recognised as a barrier to quantitative assays and has been the focus of a number of groups. Metering techniques, valve designs, and timing controls have been explored on paper to control the delivery of the dosed samples to the channels, while attempting to ensure both the low-cost nature of the material and that fabrication processes are not compromised (Chen et al. 2012; Lutz et al. 2013; Jahanshahi-Anbuhi et al. 2012; Noh and Phillips 2010b).

7.4.2 Microfluidic Flow and In-Situ Operations

As noted above, the fundamental advantage of paper-based diagnostics is the ability to transport fluid spontaneously using an inexpensive, sustainable, easily sourced material already integrated successfully into the supply chain of many established sectors such as medical technologies. Equally beneficial is the rich history of process technologies developed to shape and cut paper very precisely and also to deposit patterns of inks on to paper surfaces for text or graphical applications.

These well-known technologies are applied to the fabrication of channels or reaction zones from paper materials for transport of biological components, one of the challenges prolific in the literature. This can be simply to transport the sample or in a more complex setup to move the analyte through a multi-step process defined by digital programming by channel modification (Walsh et al. 2010). Reviewing the broad range of techniques identified in Table 7.3, it can be seen that there are four main paradigms for guiding fluid flow in paper-based diagnostics, namely:

1. Defining selective wettability of paper to provide flow direction.
2. Destruction of porosity (this includes imbibition into the pores and increasing density) to prevent or reduce the sample flow rate.
3. Direct shaping/cutting of the paper to provide the necessary boundaries and structures.
4. Using standard microfluidics to guide the sample to a paper reaction zone (not discussed in detail).

Selective Wettability

As noted in Sect. 7.4.1, paper is an ideal material for the transport of aqueous solutions and so it is a natural development to treat the cellulose fibres in order to modify them to guide the flow. Changing the surface chemistry leaves the porous channels still open and so it is critical for this approach that the capillary pressure (Eq. 7.1) is sufficiently negative as to ensure that no flow can occur beyond the channel walls. The aim is to apply a fluid that efficiently coats the surface, dries and leaves a strongly adsorbed surface layer with a very high liquid/solid interfacial energy. The chosen coating materials are normally hydrophobic as the samples are most often aqueous. Hydrophobic particles and colloids have been used to adhere to the paper and stop the capillary flow, with alkyl ketene dimer (AKD) being one of the most common of these sizing agent (Fig. 7.7). A range of polymers have also been used, including polydimethylsiloxane (PDMS) (Bruzewicz et al. 2008a), polystyrene (Abe et al. 2008) and certain photo resists (Martinez et al. 2007b) to name

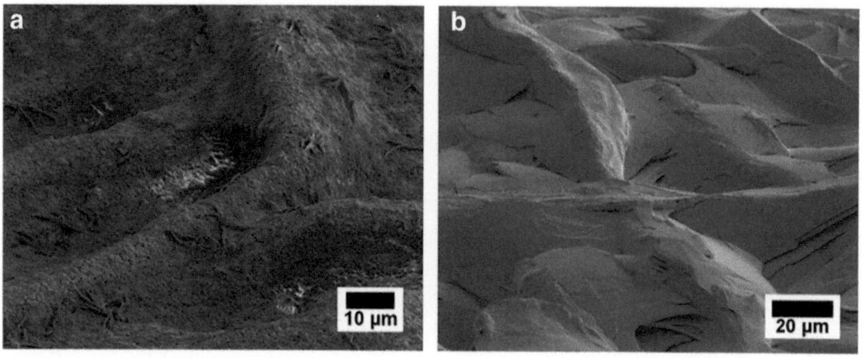

Fig. 7.7 (**a**) Whatman 1 chromatography paper coated in an excess of alkyl ketene dimer (Hydrores, Kemira) to ensure hydrophobicity; (**b**) hot-melt hydrophobic ink (SunChemical) deposited on Whatman 1 chromatography paper

but a few. Activating paper by attaching various functionalizing groups, such as modification with silanes, is a common approach but many such moieties render paper hydrophobic. To convert paper back to its natural hydrophilic nature either solvents (Abe et al. 2008) or laser treatment (Chitnis et al. 2011) can be used. To avoid these treatments and to increase the surface area for interaction with capture proteins, functionalization is performed on particles or similar surfaces and then added on to paper. Fusion 5, a commercial paper (Whatman), has larger pore size in the reaction zone so that functionalized particles can be retained and act as a horizontal immunochromatography column.

Traditional techniques for pattern transfer on to paper (such as gravure, flexography, inkjet and laser printing) deliver inks for text and graphical applications. These are complex fluids with a range of surface tensions and viscosities, containing polymers, colloids, UV-sensitive materials and other functional components. To identify the challenges in using such printing approaches to functionalize paper for biological applications, the dissimilarities need to be examined. For example, a move to biosensing through patterning of paper requires a far more precise and robust patterning technique than for standard industrial printing. Minor nozzle misfiring and blockages in inkjet printing may be unnoticed in many graphical applications but are critical to defining channel edges and biological material concentrations. Additional technological challenges exist, including the fact that the sample will be in constant contact with the modifier material throughout the device and so must be designed to avoid non-specific binding of the analyte. In addition, while a wide range of materials have been shown to work in the literature, the long-term stability of these materials is critical to understand. For example, alkyl ketene dimer is known to continue to spread over a long duration, upon drying of the carrier fluid used to coat the paper, changing its hydrophobic behaviour.

Closure of Porosity

Modification of porosity was part of the initial breakthrough in paper-based diagnostics as it is an inherent side-effect of wax printing. The proliferation of reports relating to wax printing is due to the realisation that subsequent melting of the wax allowed complete imbibition into the pores followed by re-solidification to ensure excellent channel sealing. In effect, removing open pores leads to densification of the paper. Increasing density by addition of SU-8 photopolymer or wax are now standard approaches reported in the literature pioneered by the Whiteside's group. However, the lithographic technique (Martinez et al. 2009) is not inexpensive, requires multiple steps and generates significant volumes of waste material. Often the materials used leave solvent residue (e.g. toluene, acetone, SU8 developer) which would be expected to affect biological components Waxes imbibe easily into the material, but careful process design is required to ensure full penetration into the paper without obscuring the channels. Significant volumes of such materials are required as paper is usually ~70 % porous and the majority of this volume must be filled to prevent flow. This leads to concerns over compromising the benefits of paper with the introduction of large quantities of the non-cellulose materials into the device.

Densification of paper and flow control can also be achieved both through the method of paper fabrication and by the careful selection of the specific cellulose fibres. This can again be used to modify or tune the flow rates for a specific application. This approach shows an initial benefit for flow delay but to allow flow guiding or channeling, the paper must be compressed so severely that the deformation is similar to that caused by die-cutting, as shown in Fig. 7.8.

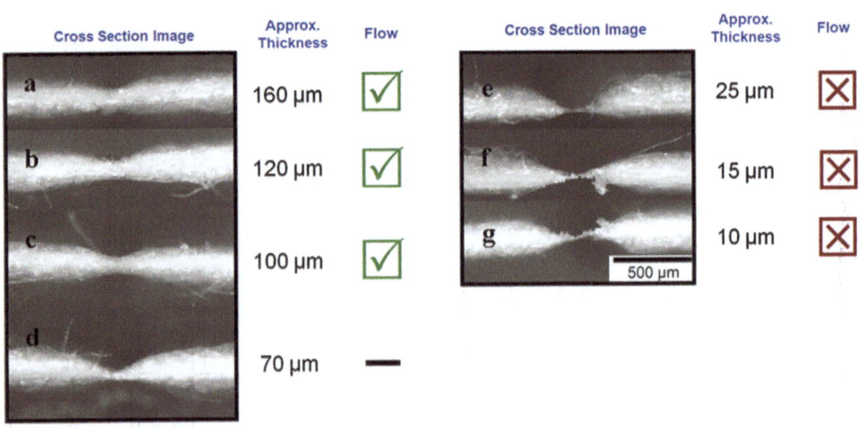

Fig. 7.8 A 25 mm × 0.44 mm carbon steel wire was used as a die to emboss Whatman 1 chromatography paper with loads of (**a**) 200 N, (**b**) 500 N, (**c**) 1,000 N, (**d**) 1,500 N, (**e**) 2,000 N, (**f**) 2,500 N, (**g**) 3,000 N. With increasing levels of compression, the final paper structure decreases in thickness and prevents capillary flow. The approximate thickness of the thinnest point of the cross section is shown, along with an indication of capillary flow. A *green tick* indicates that capillary flow of water still occurs across the embossed region. A *black line* indicates a temporary blocking with subsequent flow. A *red cross* signifies effective blocking of capillary wicking

Direct Shaping and Cutting

Significant work has been carried out by simply cutting paper into a channel or reaction zone design. Traditionally this approach is used for 96-well plate assays with polycarbonate or polyethersulfone discs. With the ease of use of laser cutting, die-cutting or perforation, complex patterns are also achievable that allow specific flow programming to be carried out. Examples of channel dimensions and junctions are shown in Fig. 7.9a–b. As discussed by Rezk et al. (2012) accurately designed paper junctions can be created to allow an understanding of flow, mixing and reactivity.

The future of paper-based diagnostics is most likely to be found in this more fundamental work, where paper could be used in quantitative multi-step processes. The rapid progression of this approach can be seen in the work of the Yager's group (Fu et al. 2012; Lutz et al. 2011) where laser-cut paper devices are used to control autonomous sequential fluid delivery, an important step towards multi-step processes. In addition, this approach is incorporated into a device for the detection of the malaria protein *Pf*HRP2. In this case the programmed approach delivers the labelled antibody, wash step and amplification reagent to the capture area in a significant step towards a commercial multi-step device.

Common Challenges

A range of additional, common challenges face the implementation of paper-based channels using any of the techniques described above. First, while paper is chosen because of its natural ability to transport fluid, it is still vulnerable to the same challenge as all microfluidic approaches, namely the difficulty in ensuring good mixing within a channel. Fluids flowing through sub-millimetre scale channels naturally have very low Reynolds numbers and exhibit laminar flow. When sample flow is required to pick up, mix or react with biological reagents it involves only diffusion-controlled mixing, and so the homogeneity of the resulting solution is difficult to predict. Recent work has focused on this challenge for paper channels and shown that for potential multi-step processes an additional stimulus to mixing may be required, such as, for example, by the use of surface acoustic waves (Rezk et al. 2012).

In addition to this challenge, paper lacks one of the key properties required of microfluidics, which is a sealant to prevent the sample from contacting external surfaces. This leads to three challenges:

1. Any aqueous sample open to the atmosphere will evaporate. For slowly flowing tests or small channels with a high ratio of outer surface area to bulk volume, this can lead to changes in the analyte concentration with distance from the dosing point.
2. Where fluid flow through paper is required, the fluid front moves within the paper matrix. Upstream it tends to saturate the paper quickly and in fact can

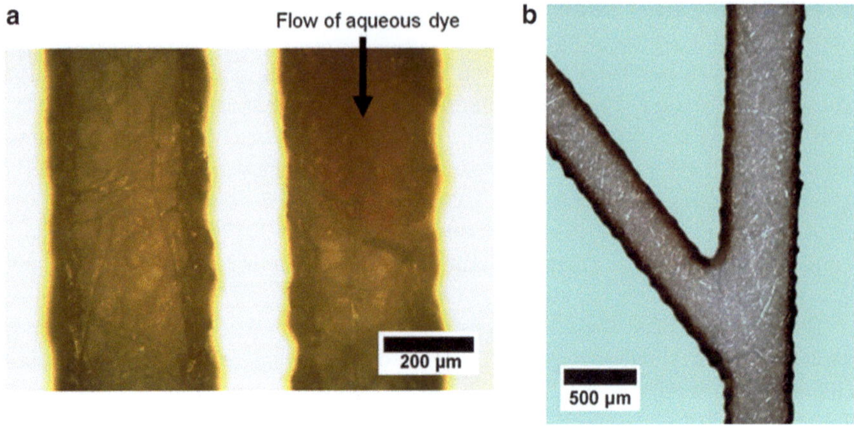

Fig. 7.9 (**a**) A device was fabricated where a single sample dosing region splits into narrow, parallel paper channels. (**b**) A y-shaped junction of such a device is shown from a complex device, fabricated using laser-cutting

Fig. 7.10 A laser-cut paper channel (approximately 1 mm in diameter) is placed on a glass microscope slide with the main aqueous sample flow moving from right to left. A T-junction is positioned in the channel with dried dye at one end. The flow reaches the point where the dye is positioned and returns along a secondary capillary channel against the bulk flow

displace fluid to the outer surface of the paper, leading to additional routes of diffusion and flow (Bico and Quéré 2003).

3. Maintenance of sufficient mechanical stability is quite difficult to achieve. With the exception of wax printing or SU-8 lithography, there is very little structural support for stand-alone paper surfaces. Samples that are delivered with the paper lying flat on a supporting surface can experience a secondary capillary flow observed between the channel material and the support as shown in Fig. 7.10.

This can give significant flow both with and against the flow direction of the bulk of the sample. Buckling and warping may also be problematic, leading to cross-channel contamination.

Recent work (Fu et al. 2012) uses paper with a plastic cover. However, it must be considered to what extent this additional packaging and the necessary assembly step impacts on the benefits of paper-based diagnostics. It has been shown that with careful construction, the plastic support can be used to control problematic secondary capillary flow and can help to tune the flow rate within the paper matrix. This work was stimulated by another challenge inherent in the use of paper for transporting samples, namely its heterogeneity in density, pore size distribution, anisotropy, structural components and the natural variation in surface chemistry. Techniques have been developed to both record and attempt to improve these variations (Roberts 1996; Lappalainen et al. 2010), but there is a limit to the improvement feasible in paper formation and heterogeneity, due to the nature of its components and fabrication.

Capture of Analyte and Signal Generation

This third operation occurs within the porous paper matrix and is the key functional step. For paper-based diagnostics to fulfil their purpose, test methods must be translated to this medium. The next section reviews detection techniques in lab-on-paper devices.

7.4.3 Detection Techniques for Lab-on-Paper Devices

Lab-on-a-chip paper-based microfluidic analytical devices have shown potential for use in myriad of applications. Paper-based devices have been fabricated with a wide variety of reaction and sensing mechanisms for specific applications. The ambitious goal of these devices is to prepare devices on paper as an alternative to plastic and glass platforms. Here, the main reaction and sensing mechanisms of paper-based devices are presented and critically discussed, together with an evaluation of their success in practical, real-world applications.

Paper-Based Microfluidic Devices for Sensing

Paper-based devices have been utilised for analytical and biochemical analysis, including lateral flow assays (LFAs), dipstick assays, and microfluidic paper-based devices. Paper-based devices for diagnostics have been used for the determination of pH, blood glucose, urine analytes, heavy metals, ions, hormones, infectious diseases, among others. In terms of reaction mechanism, these paper-based devices may be categorised as using optical (e.g. colorimetric chemical or biological reactions), electrochemical or hybrid sensing systems.

Colorimetric Paper-Based Sensing Devices

Colorimetric paper-based diagnostic devices are the most widely used type of paper-based optical sensors. Traditionally, colorimetric paper-based devices are utilised for "yes/no" semi-quantitative assays with the help of a calibration chart. The change in colour of the assay, due to a chemical or enzymatic reaction, may be visualised and interpreted by the naked eye, which makes this technique simple and attractive for rapid and practical applications. The chemical or enzymatic reactions take place in defined reaction zones that are used to capture various target analytes in the same device. Microfluidic channels are generally fabricated with hydrophobic patterns, for example utilising wax. Reaction zones are functionalized by spotting the required reagents within these zones. The colorimetric reaction occurs once the target analytes react with the previously immobilised reagents within the reaction zones, leading to a corresponding colour change (Martinez et al. 2007a). Martinez et al. (2007a) in the Whitesides' Group at Harvard University in the USA presented the first work on colorimetric paper-based devices for the colorimetric determination of glucose oxidase and protein. Their device consisted of paper printed with photoresist to define microfluidic channels, and reaction zones containing glucose oxidase, horseradish peroxidase and potassium iodide for glucose detection and tetrabromophenol blue for protein detection, in this case for bovine serum albumin: (BSA). Regarding the glucose assay, the presence of glucose resulted in a colour shift from clear to brown as a result of the enzymatic oxidation of iodide to iodine. For the BSA protein assay, the protein's presence led to a colour change of tetrabromophenol blue from yellow to blue. The detection range of such devices was reported as 2.5–500 mM for glucose and 0.38–75 μM for BSA (Fig. 7.11) (Martinez et al. 2007a). Other colorimetric devices have been fabricated by cutting chromatography paper in a tree-shaped format, modified with bromophenol blue, for the simultaneous calibration and semi-quantitative determination of BSA in artificial urine (Wang et al. 2010).

Dungchai and collaborators have developed microfluidic paper-based devices containing multiple colorimetric indicators to increase the accuracy of these types of assays, by increasing the ability to better discriminate differences among colours. Reaction zones in this device were modified with 4-aminoantipyrine, 3,5-dichloro-2-hydroxy-benzenesulfonic acid, *o*-dianisidine dihydrochloride, potassium iodide, acid yellow and acid black to quantify glucose (0.5–20 mM), uric acid (0.1–7 mM) and lactate (1–25 mM). The use of a multiple colorimetric indicators gave an accuracy of >90 %, while a single indicator standard device had ~70 % accuracy (Dungchai et al. 2010). Other glucose sensing paper devices have been constructed with cerium (IV) oxide nanoparticles, chitosan and glucose oxidase (Ornatska et al. 2011). More recently, paper-based devices in which the test zone have been functionalised with silver nanoparticles, homocysteine and dithiothreitol, were able

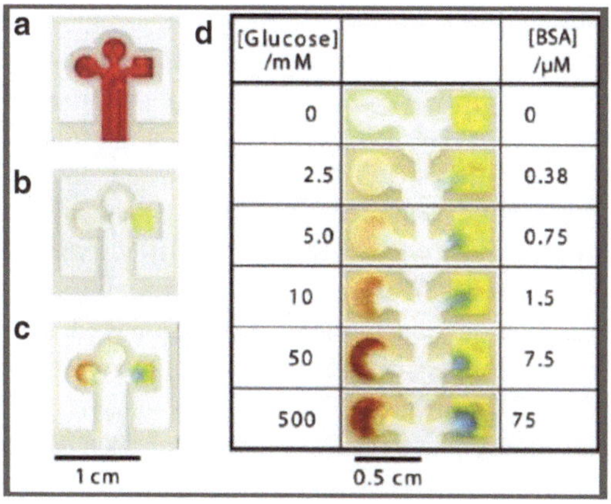

Fig. 7.11 μPADs for analysis of glucose and urine developed by Whitesides Group. (**a**) 5 μl of waterman red ink is distributed in patterned paper to check the integrity of the paper. (**b**) The square zone on the right is for protein test while the circular zone on the left is for testing glucose. (**c**) Assay is shown with positive assays for glucose (20 mM) and protein (75 μM) using 5 μl of artificial urine solution. (**d**) It shows the assay with range of concentrations

to detect copper, due to colour change (Ratnarathorn et al. 2012). Through visual inspection, this assay presented a detection limit of ~7.8 nM for copper. Paper-based devices also have been used to detect glucose and ketones in artificial urine and nitrite in saliva samples. Here, the reactions zones were modified with potassium iodide to detect glucose, while *N*-(1-napthyl)ethylenediamine was employed to detect nitrite. Ketones were detected through an assay consisting of a sample spotting zone, followed by a middle pretreatment zone and a final reaction zone. For this assay, an acetoacetate sample reacts with pre-deposited glycine located in the middle pretreatment zone, leading to the formation of an imine. This imine subsequently flows to the final detection zone and reacts with sodium nitroprusside, resulting in a magenta colour (Klasner et al. 2010).

Colorimetric DNase I and adenosine aptamer paper-based tests have also been developed. For this particular DNase I test, gold nanoparticles are initially crosslinked with DNA chains. The presence of the enzyme DNase I lead to the enzymatic degradation of the DNA chains, leading to a dispersion of gold nanoparticles, acting as colorimetric probes and a corresponding colour change to red. Similarly, gold nanoparticles may be crosslinked to specific aptamer, in this case an adenosine aptamer. In the presence of adenosine, the aptamer dissociates from the gold nanoparticles and binds to adenosine. This leads to aggregation of the gold nanoparticles, which also results in a bright red colour (Zhao et al. 2008). Other colorimetric assays are capable of detecting enzyme inhibitors or the products of enzyme activity. These assays are based on the reaction of enzymatic products with functionalized test zones leading to colour change.

Colorimetric enzyme-linked immunosorbent assays (ELISA), which are tests that combine the specificity of antibodies with specific enzymatic catalysis to provide specific and sensitive assays for routine measurements of a myriad of analytes in patient samples, are ubiquitous in research, clinical and medical laboratories worldwide. Recently, ELISA which is traditionally performed in 96-microzone plastic plates has been fabricated on 96-microzone paper plates (paper-based ELISA or P-ELISA). The paper plates are fabricated by patterning microzones with hydrophobic polymer walls on hydrophilic paper. P-ELISA combines the sensitivity and specificity of traditional ELISA with the simplicity, low cost and ease-of-use attributes of paper-based platforms. Currently, P-ELISA has shown to be faster and less expensive than conventional ELISA, but assay results have been found to be less sensitive and further paper platform development is ongoing (Cheng et al. 2010).

Lateral flow colorimetric paper-based assays have been developed to provide a simple and general one-step signal amplification strategy for the detection of nucleic acid sequences (Hu et al. 2013). Here oligonucleotide-linked gold nanoparticle (AuNP) aggregates were utilised in order to enhance the sensitivity of the nucleic acid lateral flow assays. A nucleic acid sequence of the human immunodeficiency virus type 1 (HIV-1) was used as the model analyte, and the colorimetric sensing assay was capable of achieving a detection limit of 0.1 nM. The same amplification strategy can be employed to enhance the sensitivity in detecting various proteins and other biomolecules on a lateral flow test strip, by simply using the desirable detector probes to the specific antibodies or aptamers of interest.

There are many other studies on colorimetric detection of analytes for a variety of applications, including medical diagnostics, environmental monitoring and health and food safety. It is important to note that despite their simplicity, colorimetric sensing in paper-based devices generally presents a disadvantage in terms of the colour inhomogeneity within the reaction zones, which poses a challenge for accurate colour analysis. In addition, paper-based devices may exhibit background noise from the paper or sample. For example, colorimetric assays to detect the presence of a specific analyte in whole blood samples often require a separation step or blood separator device. Despite the above challenges, the use of UV–Vis spectrophotometers, scanners, handheld readers and mobile phone digital cameras, with sophisticated algorithms for ambient light compensation, together with colour calibration, processing and analysis, holds great promise to address the challenges associated with colour variations in colorimetric paper-based assays (Martinez et al. 2008a).

Fluorescence Paper-Based Sensing Devices

The basic mechanism used in fluorescence sensing devices is the detection of a signal resulting from interaction between the target and fluorescent molecules, nanoparticles and nanomaterials, detected through optical-electronic methods. Fluorescence paper-based sensing devices were first demonstrated on paper microzone plates. Paper-based fluorescence assays often present a few challenges since many commercially available paper substrates also self-fluoresce (autofluorescence) due to additives that are incorporated within their structure to increase their whiteness, leading to high background signal interference (Pelton 2009b). In comparison to colorimetric assays, fluorescence detection also often requires additional sensing instrumentation. Early studies of paper-based fluorescence sensors demonstrated that 12.5 pmol of fluorescein isothiocyanate-labelled bovine serum albumin generated a relative fluorescence of 2,700 ± 850 (AU) (Carrilho et al. 2009b). These fluorescence assays showed concentration sensitivity comparable to plastic microzone plates, but the average relative standard deviations were found to be higher. These sensing limitations were primarily attributed to the scattering of light by the cellulose fibres and the influence of the refractive index difference between the air and the fibres. Other paper-based fluorescence sensing platforms have utilised DNA-conjugated microgels for the detection of DNA (Ali et al. 2009). Here, the detection was performed in the following steps: (1) targeting DNA promoted ligation of a DNA primer to the microgel-bound DNA, (2) rolling circle amplification (RCA) between the primer and a circle DNA, and (3) hybridization of the RCA products and fluorescent DNA probe. Fluorescent sensing paper devices have also utilised non-enzymatic nucleic acid circuits based on strand-exchange reaction for the detection of target sequences (Allen et al. 2012).

Recently, researchers have developed a multiplexed solid-phase nucleic acid hybridization assay on a paper platform, using multicolour immobilised quantum dots (QDs) as the donors in fluorescence resonance energy transfer (FRET) sensing (Noor and Krull 2013; Noor et al. 2013). In the construction of these devices, paper was functionalized with imidazole groups in order to immobilise two different types of QD-probe oligonucleotide conjugates. The QDs used were green-emitting (gQDs) and red-emitting (rQDs), which served as donors with Cy3 and Alexa Fluor 647 acceptors, respectively. The gQD/Cy3 FRET pair acted as the internal standard, while the rQD/Alexa Fluor 647 FRET pair was used as the detection channel, combining the analytical test areas and control in the same sensing device. Here, the hybridisation of a dye-labelled oligonucleotide target provides the proximity for the FRET sensitised emission from the acceptor dyes, which is used to detect the analytical signal. In the multicolour format used, these hybridisation assays had a detection limit of 90 fmol and an upper dynamic range limit of approximately 3.5 pmol. In this study, the two-plex hybridization assay selectivity was shown through the detection of a single nucleotide polymorphism at a contrast ratio of 50:1 (Noor and Krull 2013).

Chemiluminescent Paper-Based Sensing Devices

Chemiluminescence is a light-emission caused by a chemical reaction in which the product has an excited intermediate capable of emitting a photon, and may be used as a sensing mechanism. Chemiluminescence sensing methods are based on the measurement of the intensity of the light emitted by the chemical reaction. For example, in the presence of reactants (e.g. luminol and hydrogen peroxide), and a catalyst or an excited intermediate (e.g. 3-aminophthalate), the chemical reaction leads to the emission of light. In this reaction, hydrogen peroxide catalyses the oxidation of luminol to 3-aminophthalate and the subsequent decay of the excited state to a lower energy level of the excited intermediate, leads to the emission of light. This light-emitting process can be enhanced through the use of modified phenols, such as piodophenol (Yetisen et al. 2013).

Chemiluminescence paper-based sensing devices have been developed to detect various analytes, including tumour biomarkers and various biological targets (Ge et al. 2012; Wang et al. 2012b; Yu et al. 2011a, b). As an example, chemiluminescence sensors have utilised oxidoreductases (glucose oxidase, urate oxidase) together with chemiluminescence reactions between the generated hydrogen peroxide and a rhodamine derivative, to detect glucose and uric acid in artificial urine samples (Yu et al. 2011a). Another example involves the use of chemiluminescence for the detection of tumour biomarkers that was developed in conjunction with an ELISA platform containing specific antibodies that were covalently bonded to the paper matrix (Ge et al. 2012). Recently, a chemiluminescence immunoassay was developed and incorporated in paper-based analytical devices (Wang et al. 2012a). This paper-based chemiluminescence assay presented rapid, stable, high-throughput and reusable capabilities in the determination of biological analytes. This chemiluminescence paper-based sensor utilised a periodate oxidation process that can covalently link polysaccharides and proteins, to activate the sensors for antibody immobilisation. This paper-based sandwich chemiluminescence immunoassay was shown to be sensitive, specific, and useful for the determination of biomarkers in human serum (Wang et al. 2012a).

Electrochemical Paper-Based Sensing Devices

Electrochemical sensing techniques are based on the measurement of electrical signals from a reaction in a particular system in order to gather information regarding the system. Here, an electrochemical component in the system serves as the principal transducing element. Generally, as the recognition elements, electrochemical detection systems employ enzymes, antibodies and nucleic acids.

Electrochemical sensing generally requires a reference electrode, a counter-electrode and a working electrode, which is traditionally referred to as the sensing or redox electrode. Electrochemical sensing techniques can be classified in six main categories (amperometric, voltammetric, conductometric, potentiometric, impedimetric and field-effect). Amperometric and voltammetric sensing are used to detect the generation of measurable currents or voltages within the system, respectively. Conductometric sensing involves the determination of varying conductive properties of a medium between the anodic and cathodic electrodes of the system. Potentiometric sensing involves the measurement of charge accumulation or a potential change within the system. Impedimetric sensing is used to measure impedance within the system. Field-effect sensing utilises transistors to measure current due to a potentiometric effect at a gate electrode. Paper-based devices have been fabricated which use all the above electrochemical techniques. One of the principal advantages of electrochemical paper-based detection, in comparison with optical methods, is the fact that they are not affected by ambient light, since the measurement of electrolytic reactions substitute optical readouts.

Several studies have demonstrated paper-based electrochemical devices that have the capability to detect multiple analytes (Dungchai et al. 2009; Nie et al. 2010; Maattanen et al. 2013). The design of electrochemical paper-based devices generally involves the use of a conventional potentiostat or a commercial reader to control the device and allow the measurement and read outs (Nie et al. 2010; Xiang and Lu 2011). Recently, in an attempt to produce stand-alone electrochemical paper devices, paper-based batteries have been developed to allow their integration with electrochemical sensing mechanisms (Hu et al. 2009; Liu and Crooks 2011). The high surface roughness and porosity in paper-based devices provide a large surface area for the deposition of electrodes and sensing receptors, thereby generally improving the response of the device (Sarfraz et al. 2012). On the other hand, a high roughness in paper-based devices poses challenges in the physical deposition steps of the various materials used in device fabrication (Hu et al. 2012). To tackle this challenge, electrochemical sensors have increasingly utilised expertise from microelectronics and printing research areas to improve electrode, battery and circuitry deposition.

Electrochemiluminescent Paper-Based Sensing Devices

Electrochemiluminescence (ECL), which combines the advantages of chemiluminescence and electrochemistry, continues to impact diverse areas ranging from chemical analysis to the molecular-level understanding of biological processes (Richter 2004). Electrochemiluminescence sensing presents higher sensitivity and a wider dynamic concentration response range in comparison to traditional electrochemical and chemiluminescent assays. Other advantages of electrochemiluminescence sensing include capability of being both potentially and spatially controlled. The use of ECL methods of sensing on paper-based devices has opened up many novels possibilities for the detection of chemically and biologically relevant targets (Delaney et al. 2011). Recently, ECL sensing has been utilised in an immunoassay paper-based microfluidic format (Ge et al. 2012). The electrodes for this device were screen-printed on paper and a wax-patterned three-dimensional paper-based ECL device was prepared. The ECL paper-based sensor utilised the typical tris-(bipyridine)-ruthenium(II)-tri-*n*-propylamine ECL system. The performance of the 3D paper-based ECL device was tested through the diagnosis of four tumour markers in real clinical serum samples. By utilising a simple device-holder and a section-switch assembled directly on the analyser, eight working electrodes were sequentially placed into the circuit to trigger the ECL reaction in the sweeping range from 0.5 to 1.1 V at room temperature. Furthermore, this 3D paper-based ECL device can be easily integrated with current and emerging paper electronics to further develop simple, sensitive, low-cost, disposable and portable paper-based devices for a myriad of applications (Ge et al. 2012).

7.4.4 Future Perspectives and Conclusions

There is no doubt that paper-based microfluidics have progressed a long way from the position a decade ago. However, to have any equitable view on the future development of the paper-based diagnostics one should not ignore similar progress in competing platforms. Most of them are at a similar stage of development and the race to the commercial success will be determined by how well a particular test can capitalise on the benefits of the material used. Any platforms that failed to deliver a marketable product in the coming decade or so would never be able to move beyond cordons of academia and the extended research world. There are three interrelated market demands for all the platforms: low cost, better quantification and rapid testing. It still remains a challenge to see all these characteristics combined in one product. The importance of low-cost is discussed earlier in Sect. 7.1.3, but better quantification in a given time remains a major challenge. If a test cannot deliver results within 5 min or so then its relevance for clinicians may be limited. Clinicians usually compare quality of these tests against centralised laboratory equipment, which is highly quantitative with high throughput e.g. Abbott's ARCHITECT c16000 performs 1,800 tests per hour using 15 μl of sample. However, even with the current state of development low-cost diagnostics can find a niche. There is a need of qualitative screening tests for large populations for various disease indications. Such markets present a huge potential for paper-based diagnostics even at their current level of technology readiness.

Acknowledgments The research leading to these results has received funding from the European Union Seventh Framework Programme (FP7/2007-2013) under grant agreement number 263061.

References

(1979). US Patent 4160008
(1982). US Patent 4313734
(1983). US Patent 4376110
(1984). US Patent 4435504
(1987). US Patent 4703017
(1989). US Patent 4855240
(1990). US Patent 4954452
(1991). US Patent 4983416
(1997). European Patent 0810436A1
Abe K, Suzuki K, Citterio D (2008) Inkjet-printed microfluidic multianalyte chemical sensing paper. Anal Chem 80(18):6928–6934
Ali MM, Aguirre SD, Xu YQ, Filipe CDM, Pelton R, Li YF (2009) Detection of DNA using bioactive paper strips. Chem Commun 43:6640–6642
Alkasir RSJ, Ornatska M, Andreescu S (2012) Colorimetric paper bioassay for the detection of phenolic compounds. Anal Chem 84(22):9729–9737
Allen PB, Arshad SA, Li BL, Chen X, Ellington AD (2012) DNA circuits as amplifiers for the detection of nucleic acids on a paperfluidic platform. Lab Chip 12(16):2951–2958

Araujo AC, Song YJ, Lundeberg J, Stahl PL, Brumer H (2012) Activated paper surfaces for the rapid hybridization of DNA through capillary transport. Anal Chem 84(7):3311–3317

Assembly GW (2005) Medicines and older people: implementing medicines related aspects of the national service framework for older people in Wales

Balu B, Berry AD, Hess DW, Breedveld V (2009) Patterning of superhydrophobic paper to control the mobility of micro-liter drops for two-dimensional lab-on-paper applications. Lab Chip 9 (21):3066–3075

Berson SA, Yalow RS (1959) Quantitative aspects of the reaction between insulin and insulin-binding antibody. J Clin Invest 38:1996–2016

Bhandari P, Narahari T, Dendukuri D (2011) 'Fab-Chips': a versatile, fabric-based platform for low-cost, rapid and multiplexed diagnostics. Lab Chip 11(15):2493–2499

Białopiotrowicz T, Jańczuk B (2001) Wettability and surface free energy of bovine serum albumin films. J Surfactants Deterg 4(3):287–292

Bico J, Quéré D (2003) Precursors of impregnation. Europhys Lett 61(3):348

Bracher PJ, Gupta M, Mack ET, Whitesides GM (2009) Heterogeneous films of ionotropic hydrogels fabricated from delivery templates of patterned paper. ACS Appl Mater Inter 1 (8):1807–1812

Brujan E-A (2011) Cavitation in non-Newtonian fluids Cavitation in non-Newtonian fluids by Emil Brujan. Springer, Berlin. ISBN 978-3-642-15342-61

Bruzewicz DA, Reches M, Whitesides GM (2008a) Low-cost printing of poly (dimethylsiloxane) barriers to define microchannels in paper. Anal Chem 80(9):3387–3392

Bruzewicz DA, Reches M, Whitesides GM (2008b) Low-cost printing of poly(dimethylsiloxane) barriers to define microchannels in paper. Anal Chem 80(9):3387–3392

Carrilho E, Martinez AW, Whitesides GM (2009a) Understanding wax printing: a simple micropatterning process for paper-based microfluidics. Anal Chem 81(16):7091–7095

Carrilho E, Phillips ST, Vella SJ, Martinez AW, Whitesides GM (2009b) Paper microzone plates. Anal Chem 81(15):5990–5998

Couderc S, Ducloux O, Kim BJ, Someya T (2009) A mechanical switch device made of a polyimide-coated microfibrillated cellulose sheet. J Micromech Microeng 19(5)

Cravioto A, Lanata CF, Lantagne DS, Balakrish Nair G (2011) Final report of the independent panel of experts on the cholera outbreak in Haiti

Crowther J (2009) The ELISA Guidebook Humana. Totowa, NJ

Cunha AG, Gandini A (2010) Turning polysaccharides into hydrophobic materials: a critical review. Part 1. Cellulose. Cellulose 17(5):875–889

Chen H, Cogswell J, Anagnostopoulos C, Faghri M (2012) A fluidic diode, valves, and a sequential-loading circuit fabricated on layered paper. Lab Chip 12(16):2909–2913

Chen YW, Stott K, Perutz MF (1999) Crystal structure of a dimeric chymotrypsin inhibitor 2 mutant containing an inserted glutamine repeat. PNAS 96(4):12571

Cheng CM, Martinez AW, Gong JL, Mace CR, Phillips ST, Carrilho E, Mirica KA, Whitesides GM (2010) Paper-based ELISA. Angew Chem Int 49(28):4771–4774

Chitnis G, Ding Z, Chang C-L, Savran CA, Ziaie B (2011) Laser-treated hydrophobic paper: an inexpensive microfluidic platform. Lab Chip 11(6):1161–1165

Delaney JL, Hogan CF, Tian JF, Shen W (2011) Electrogenerated chemiluminescence detection in paper-based microfluidic sensors. Anal Chem 83(4):1300–1306

Dieterich K (1902) Testing paper and method of making same. US Patent 691249

Dungchai W, Chailapakul O, Henry CS (2009) Electrochemical detection for paper-based microfluidics. Anal Chem 81(14):5821–5826

Dungchai W, Chailapakul O, Henry CS (2010) Use of multiple colorimetric indicators for paper-based microfluidic devices. Anal Chim Acta 674(2):227–233

Dungchai W, Chailapakul O, Henry CS (2011) A low-cost, simple, and rapid fabrication method for paper-based microfluidics using wax screen-printing. Analyst 136(1):77–82

Eggins BR (2008) Chemical sensors and biosensors, vol 28. Wiley, Somerset, NJ

Finkbeiner S (2010) Bridging the valley of death of therapeutics for neurodegeneration. Nat Med 16(11):1227–1232

Free AH, Adams EC, Kercher ML, Free HM, Cook MH (1957) Simple specific test for urine glucose. Clin Chem 3(3):163–168

Fu E, Liang T, Spicar-Mihalic P, Houghtaling J, Ramachandran S, Yager P (2012) Two-dimensional paper network format that enables simple multistep assays for use in low-resource settings in the context of malaria antigen detection. Anal Chem 84 (10):4574–4579

Gao KZ, Shao ZQ, Wu X, Wang X, Li J, Zhang YH, Wang WJ, Wang FJ (2013) Cellulose nanofibers/reduced graphene oxide flexible transparent conductive paper. Carbohyd Polym 97 (1):243–251

Ge L, Yan JX, Song XR, Yan M, Ge SG, Yu JH (2012) Three-dimensional paper-based electrochemiluminescence immunodevice for multiplexed measurement of biomarkers and point-of-care testing. Biomaterials 33(4):1024–1031

Goldberg DA (1980) Isolation and partial characterization of the Drosophila alcohol dehydrogenase gene. Proc Natl Acad Sci U S A 77(10):5794–5798

Gu Z, Zhao MX, Sheng YW, Bentolila LA, Tang Y (2011) Detection of mercury ion by infrared fluorescent protein and its hydrogel-based paper assay. Anal Chem 83(6):2324–2329

Hagist C, Kotlikoff L (2005) Who's Going Broke? Comparing Growth in Healthcare Costs in Ten OECD. Countries, National Bureau of Economic Research

Hawkes R, Niday E, Gordon J (1982) A dot-immunobinding assay for monoclonal and other antibodies. Anal Biochem 119(1):142–147

He QH, Ma CC, Hu XQ, Chen HW (2013) Method for fabrication of paper-based microfluidic devices by alkylsilane self-assembling and UV/O-3-patterning. Anal Chem 85(3):1327–1331

Hodgson KT, Berg JC (1988) The effect of surfactants on wicking flow in fiber networks. J Colloid Interface Sci 121(1):22–31

Hossain SMZ, Luckham RE, McFadden MJ, Brennan JD (2009) Reagentless bidirectional lateral flow bioactive paper sensors for detection of pesticides in beverage and food samples. Anal Chem 81(21):9055–9064

Hossain SZ, Ozimok C, Sicard C, Aguirre SD, Ali MM, Li Y, Brennan JD (2012) Multiplexed paper test strip for quantitative bacterial detection. Anal Bioanal Chem 403(6):1567–1576

http://www.surface-tension.de/solid-surface-energy.htm. Accessed Jan 2013

Hu CG, Bai XY, Wang YK, Jin W, Zhang X, Hu SS (2012) Inkjet printing of nanoporous gold electrode arrays on cellulose membranes for high-sensitive paper-like electrochemical oxygen sensors using ionic liquid electrolytes. Anal Chem 84(8):3745–3750

Hu J, Wang L, Li F, Han YL, Lin M, Lu TJ, Xu F (2013) Oligonucleotide-linked gold nanoparticle aggregates for enhanced sensitivity in lateral flow assays. Lab Chip 13(22):4352–4357

Hu J, Wang S, Wang L, Li F, Pingguan-Murphy B, Lu TJ, Xu F (2014) Advances in paper-based point-of-care diagnostics. Biosens Bioelectron 54:585–597

Hu LB, Choi JW, Yang Y, Jeong S, La Mantia F, Cui LF, Cui Y (2009) Highly conductive paper for energy-storage devices. Proc Natl Acad Sci U S A 106(51):21490–21494

Hyväluoma J, Raiskinmäki P, Jäsberg A, Koponen A, Kataja M, Timonen J (2006) Simulation of liquid penetration in paper. Phys Rev E 73(3):036705

Jagadeesan KK, Kumar S, Sumana G (2012) Application of conducting paper for selective detection of troponin. Electrochem Commun 20:71–74

Jaganathan S, Vahedi Tafreshi H, Pourdeyhimi B (2008) Modeling liquid porosimetry in modeled and imaged 3-D fibrous microstructures. J Colloid Interface Sci 326(1):166–175

Jahanshahi-Anbuhi S, Chavan P, Sicard C, Leung V, Hossain SZ, Pelton R, Brennan JD, Filipe CD (2012) Creating fast flow channels in paper fluidic devices to control timing of sequential reactions. Lab Chip 12(23):5079–5085

Johnson JL (1967) Microchemical techniques in solving industrial problems. Microchim Acta 55 (4):756–762

Kim J, Wang NG, Chen Y, Lee SK, Yun GY (2007) Electroactive-paper actuator made with cellulose/NaOH/urea and sodium alginate. Cellulose 14(3):217–223

Klasner SA, Price AK, Hoeman KW, Wilson RS, Bell KJ, Culbertson CT (2010) Paper-based microfluidic devices for analysis of clinically relevant analytes present in urine and saliva. Anal Bioanal Chem 397(5):1821–1829

Kouisni L, Rochefort D (2009) Confocal microscopy study of polymer microcapsules for enzyme immobilisation in paper substrates. J Appl Polym Sci 111(1):1–10

Lankelma J, Nie ZH, Carrilho E, Whitesides GM (2012) Paper-based analytical device for electrochemical flow-injection analysis of glucose in urine. Anal Chem 84(9):4147–4152

Lappalainen T, Teerinen T, Vento P, Hakalahti L, Erho T (2010) Cellulose as a novel substrate for lateral flow assay. Nordic Pulp Paper Res J 25(4):536–550

Lei KF, Yang S-I (2013) Bundled carbon nanotube-based sensor on paper-based microfluidic device. J Nanosci Nanotechnol 13(10):6917–6923

Li C-z, Vandenberg K, Prabhulkar S, Zhu X, Schneper L, Methee K, Rosser CJ, Almeide E (2011) Paper based point-of-care testing disc for multiplex whole cell bacteria analysis. Biosens Bioelectron 26(11):4342–4348

Li MS, Tian JF, Al-Tamimi M, Shen W (2012a) Paper-based blood typing device that reports patient's blood type "in writing". Angew Chem Int Ed 51(22):5497–5501

Li X, Ballerini DR, Shen W (2012b) A perspective on paper-based microfluidics: current status and future trends. Biomicrofluidics 6:011301

Liu H, Crooks RM (2011) Three-dimensional paper microfluidic devices assembled using the principles of origami. J Am Chem Soc 133(44):17564–17566

Liu H, Xiang Y, Lu Y, Crooks RM (2012) Aptamer-based origami paper analytical device for electrochemical detection of adenosine. Angew Chem Int Ed 51(28):6925–6928

Lutz B, Liang T, Fu E, Ramachandran S, Kauffman P, Yager P (2013) Dissolvable fluidic time delays for programming multi-step assays in instrument-free paper diagnostics. Lab Chip 13 (14):2840–2847

Lutz BR, Trinh P, Ball C, Fu E, Yager P (2011) Two-dimensional paper networks: programmable fluidic disconnects for multi-step processes in shaped paper. Lab Chip 11(24):4274–4278

Maattanen A, Vanamo U, Ihalainen P, Pulkkinen P, Tenhu H, Bobacka J, Peltonen J (2013) A low-cost paper-based inkjet-printed platform for electrochemical analyses. Sensor Actuat B Chem 177:153–162

Martinez AW (2011) Microfluidic paper-based analytical devices: from POCKET to paper-based ELISA. Bioanalysis 3(23):2589–2592

Martinez AW, Phillips ST, Butte MJ, Whitesides GM (2007a) Patterned paper as a platform for inexpensive, low-volume, portable bioassays. Angew Chem Int Ed 46(8):1318–1320

Martinez AW, Phillips ST, Butte MJ, Whitesides GM (2007b) Patterned paper as a platform for inexpensive, low volume, portable bioassays. Angew Chem Int Ed 46(8):1318–1320

Martinez AW, Phillips ST, Carrilho E, Thomas SW, Sindi H, Whitesides GM (2008a) Simple telemedicine for developing regions: camera phones and paper-based microfluidic devices for real-time, off-site diagnosis. Anal Chem 80(10):3699–3707

Martinez AW, Phillips ST, Whitesides GM (2008b) Three-dimensional microfluidic devices fabricated in layered paper and tape. Proc Natl Acad Sci U S A 105(50):19606–19611

Martinez AW, Phillips ST, Whitesides GM, Carrilho E (2009) Diagnostics for the developing world: microfluidic paper-based analytical devices. Anal Chem 82(1):3–10

Müller RH, Clegg DL (1949) Automatic paper chromatography. Anal Chem 21(9):1123–1125

Mullins BJ, Braddock RD, Kasper G (2007) Capillarity in fibrous filter media: relationship to filter properties. Chem Eng Sci 62(22):6191–6198

Nery EW, Kubota LT (2013) Sensing approaches on paper-based devices: a review. Anal Bioanal Chem 405(24):7573–7595

Nie ZH, Nijhuis CA, Gong JL, Chen X, Kumachev A, Martinez AW, Narovlyansky M, Whitesides GM (2010) Electrochemical sensing in paper-based microfluidic devices. Lab Chip 10 (4):477–483

Niño MRR, Patino JR (1998) Surface tension of bovine serum albumin and tween 20 at the air-aqueous interface. J Am Oil Chem Soc 75(10):1241–1248
Noh H, Phillips ST (2010a) Fluidic timers for time-dependent, point-of-care assays on paper. Anal Chem 82(19):8071–8078
Noh H, Phillips ST (2010b) Metering the capillary-driven flow of fluids in paper-based microfluidic devices. Anal Chem 82(10):4181–4187
Noor MO, Krull UJ (2013) Paper-based solid-phase multiplexed nucleic acid hybridization assay with tunable dynamic range using immobilized quantum dots as donors in fluorescence resonance energy transfer. Anal Chem 85(15):7502–7511
Noor MO, Shahmuradyan A, Krull UJ (2013) Paper-based solid-phase nucleic acid hybridization assay using immobilized quantum dots as donors in fluorescence resonance energy transfer. Anal Chem 85(3):1860–1867
Ordway JM, Tallaksen-Greene S, Gutekunst CA, Bernstein EM, Cearley JA, Wiener HW, Dure LS (1997) Ectopically expressed CAG repeats cause intranuclear inclusions and a progressive late onset neurological phenotype in the mouse. Cell 91(6):753–763
Organization WH (2002) WHO global strategy for food safety: safer food for better health. The Organization, Geneva
Ornatska M, Sharpe E, Andreescu D, Andreescu S (2011) Paper bioassay based on ceria nanoparticles as colorimetric probes. Anal Chem 83(11):4273–4280
Peeling RW, Holmes KK, Mabey D, Ronald A (2006) Rapid tests for sexually transmitted infections (STIs): the way forward. Sex Transm Infect 82(suppl 5):v1–v6
Pelton R (2009a) Bioactive paper provides a low-cost platform for diagnostics. TrAC Trends Anal Chem 28(8):925–942
Pelton R (2009b) Bioactive paper provides a low-cost platform for diagnostics. TrAC Trend Anal Chem 28(8):925–942
Plotz CM, Singer JM (1956) The latex fixation test. I Application to the serologic diagnosis of rheumatoid arthritis. Am J Med 21(6):888–892
Pollock NR, Rolland JP, Kumar S, Beattie PD, Jain S, Noubary F, Wong VL, Pohlmann RA, Ryan US, Whitesides GM (2012) A paper-based multiplexed transaminase test for low-cost, point-of-care liver function testing. Sci Transl Med 4(152ra129):152–152ra129
PriceWaterHouseCoopers P (2007) 2020: The vision. Which path will you take
Prionics. http://www.prionics.com/diseases-solutions/tse/Prionics-Check_PrioSTRIP/. Accessed 3 Mar 2014
Ratnarathorn N, Chailapakul O, Henry CS, Dungchai W (2012) Simple silver nanoparticle colorimetric sensing for copper by paper-based devices. Talanta 99:552–557
Rezk AR, Qi A, Friend JR, Li WH, Yeo LY (2012) Uniform mixing in paper-based microfluidic systems using surface acoustic waves. Lab Chip 12(4):773–779
Richter MM (2004) Electrochemiluminescence (ECL). Chem Rev 104(6):3003–3036
Roberts JC (1996) The chemistry of paper, vol 11. Society of chemistry, Royal
Sarfraz J, Tobjork D, Osterbacka R, Linden M (2012) Low-cost hydrogen sulfide gas sensor on paper substrates: fabrication and demonstration. IEEE Sens J 12(6)
Songjaroen T, Dungchai W, Chailapakul O, Henry CS, Laiwattanapaisal W (2012) Blood separation on microfluidic paper-based analytical devices. Lab Chip 12(18):3392–3398
Southern EM (1975) Detection of specific sequences among DNA fragments separated by gel-electrophoresis. J Mol Biol 98(3):503
Stokes JR, Davies GA (2007) Viscoelasticity of human whole saliva collected after acid and mechanical stimulation. Biorheology 44(3):141–160
Su SX, Ali M, Filipe CDM, Li YF, Pelton R (2008) Microgel-based inks for paper-supported biosensing applications. Biomacromolecules 9(3):935–941
Thies W, Bleiler L (2011) 2011 Alzheimer's disease facts and figures. Alzheimer's and dementia. J Alzheimer's Assoc 7(2):208
Tian J, Kannangara D, Li X, Shen W (2010) Capillary driven low-cost V-groove microfluidic device with high sample transport efficiency. Lab Chip 10(17):2258–2264

Towbin H, Staehelin T, Gordon J (1979) Electrophoretic transfer of proteins from polyacrylamide gels to nitrocellulose sheets: procedure and some applications. Proc Natl Acad Sci U S A 76 (9):4350–4354

Tseng SC, Yu CC, Wan DH, Chen HL, Wang LA, Wu MC, Su WF, Han HC, Chen LC (2012) Eco-friendly plasmonic sensors: using the photothermal effect to prepare metal nanoparticle-containing test papers for highly sensitive colorimetric detection. Anal Chem 84 (11):5140–5145

Vaitukaitis JL, Braunstein GD, Ross GT (1972) A radioimmunoassay which specifically measures human chorionic gonadotropin in the presence of human luteinizing hormone. Am J Obstet Gynecol 113(6)

Varmus H, Klausner R, Zerhouni E, Acharya T, Daar AS, Singer PA (2003) Grand challenges in global health. Science 302(5644):398–399

Walsh P, Albin D, Gouch M (2010) Micro-channel structure method and apparatus. Patents, Google

Wang SM, Ge L, Song XR, Yan M, Ge SG, Yu JH, Zeng F (2012a) Simple and covalent fabrication of a paper device and its application in sensitive chemiluminescence immunoassay. Analyst 137(16):3821–3827

Wang SM, Ge L, Song XR, Yu JH, Ge SG, Huang JD, Zeng F (2012b) Paper-based chemiluminescence ELISA: lab-on-paper based on chitosan modified paper device and wax-screen-printing. Biosens Bioelectron 31(1):212–218

Wang W, Wu WY, Wang W, Zhu JJ (2010) Tree-shaped paper strip for semiquantitative colorimetric detection of protein with self-calibration. J Chromatogr A 1217(24):3896–3899

water.org (2014) http://water.org/water-crisis/water-facts/children/

Werner O, Quan C, Turner C, Pettersson B, Wagberg L (2010) Properties of superhydrophobic paper treated with rapid expansion of supercritical CO2 containing a crystallizing wax. Cellulose 17(1):187–198

Xiang Y, Lu Y (2011) Using personal glucose meters and functional DNA sensors to quantify a variety of analytical targets. Nat Chem 3(9):697–703

Xu MA, Bunes BR, Zang L (2011) Paper-based vapor detection of hydrogen peroxide: colorimetric sensing with tunable interface. ACS Appl Mater Inter 3(3):642–647

Yagoda H (1937) Applications of confined spot tests in analytical chemistry: preliminary paper. Ind Eng Chem Anal Ed 9(2):79–82

Yagoda H (1938) Test paper. US Patent 2129754

Yang X, Forouzan O, Brown TP, Shevkoplyas SS (2012) Integrated separation of blood plasma from whole blood for microfluidic paper-based analytical devices. Lab Chip 12(2):274–280

Yazawa I, Nukina N et al (1995) Abnormal gene product identified in hereditary dentatorubral-pallidoluysian atrophy (DRPLA) brain. Nat Genet 10:99–103

Yetisen AK, Akram MS, Lowe CR (2013) Paper-based microfluidic point-of-care diagnostic devices. Lab Chip 13(12):2210–2251

Yu JH, Ge L, Huang JD, Wang SM, Ge SG (2011a) Microfluidic paper-based chemiluminescence biosensor for simultaneous determination of glucose and uric acid. Lab Chip 11(7):1286–1291

Yu JH, Wang SM, Ge L, Ge SG (2011b) A novel chemiluminescence paper microfluidic biosensor based on enzymatic reaction for uric acid determination. Biosens Bioelectron 26(7):3284–3289

Zava TT, Kapur S, Zava DT (2013) Iodine and creatinine testing in urine dried on filter paper. Anal Chim Acta 764:64–69

Zhao WA, Ali MM, Aguirre SD, Brook MA, Li YF (2008) Paper-based bioassays using gold nanoparticle colorimetric probes. Anal Chem 80(22):8431–8437

Zigmond SH (1977) Ability of polymorphonuclear leukocytes to orient in gradients of chemotactic factors. J Cell Biol 75(2):606–616

Chapter 8
Microfluidics in Planar Microchannels: Synthesis of Chemical Compounds On-Chip

Valentina Arima, Paul Watts, and Giancarlo Pascali

Abstract Microreactors are a wide class of devices that are currently playing a prominent role in several research fields such as biology, medicine, food chemistry, environmental analysis, up to the production of compounds in organic chemistry. Several typologies of microreactors have been produced with tubular or planar shapes, of different materials and designs. In this chapter, an overview of planar microchannel-based microreactors and their application to organic chemistry is given. Initially, after recalling the main theoretical parameters of microfluidics, an introduction of the proposed technology and the main requirements to perform mixing, which is essential to perform chemical synthesis on-chip, is presented. Silicon and glass microreactors, the most common planar systems for organic chemistry, are described with the aim of pointing out the most important parameters to be taken into consideration in the planning of a specific microreactor to be used for mixing, purification or crystallization of chemicals at the microscale. Then, several applications of initially described microreactors to organic chemistry for research applications are given. In the next section, the use of planar microchannel microreactors in the field of radiochemistry is reported. The radiopharmaceutical application is not casual, being a sector in which the microreactor technology is very promising, due to the need of quickly producing small and fresh amounts of products in a controlled environment. Finally, for completeness, other approaches beyond planar microchannels are mentioned: mesoreactors towards industrial level synthesis and micro-vessels for radiochemistry.

V. Arima (✉)
CNR-NANO, Institute of Nanoscience, U.O.S. Lecce, via per Arnesano, Lecce 73100, Italy
e-mail: valentina.arima@nano.cnr.it

P. Watts
InnoVenton: NMMU Institute for Chemical Technology, Nelson Mandela Metropolitan University, P.O. Box 77 000, Port Elizabeth 6031, South Africa

G. Pascali
ANSTO Lifesciences, 67-69 Missenden Road, Camperdown, NSW 2050, Australia

J. Castillo-León, W.E. Svendsen (eds.), *Lab-on-a-Chip Devices and Micro-Total Analysis Systems*, DOI 10.1007/978-3-319-08687-3_8

8.1 Passive Mixing in Planar Microreactors

Mixing operations at the microscale in planar microreactors can be forced by external sources (as described in the theory chapter) in active mixers or be driven by mass transport phenomena in passive mixers. Parameters and general rules which govern the passive mixing will be the subject of the present discussion.

While in large scale systems mass transport is dominated by convention (due to inertial forces, an applied pressure, electromagnetic fields ...), at the microscale diffusion becomes important. In fluid dynamics there are two dimensionless parameters that describe the mass transfer processes. The balance between inertial and viscous forces is expressed by the Reynolds number, Re, which presents a critical value between laminar (<2,300) and turbulent flow (>2,300).

$$Re = \frac{Ud}{v} \tag{8.1}$$

U, d, and v denote the average velocity, the hydraulic diameter or the transverse diffusion distance, and the kinematic viscosity.

A balance between the mass transport due to convection and diffusion is expressed by the Peclet number, Pe. D represents the diffusion coefficient.

$$Pe = \frac{Ud}{D} \tag{8.2}$$

In a microreactor, most flows are laminar (low Re) without turbulence; in laminar regime, convective transport mechanisms (high Pe) limit mixing. For synthesis of chemical compounds on-chip, some degree of mixing is necessary, therefore, in order to improve mixing it is crucial to increase diffusion. Beyond the diffusion coefficient which depends on the solute dimensions and the temperature, the diffusion flow (D_f) depends on the interfacial surface area (A) and the concentration gradient (∇C). In the "art of mixing" at microscale a maximization of the parameters A and ∇C allows enhanched diffusion and improved mixing yields.

An increase of the interfacial surface area may occur by creating secondary flow patterns superimposed to the main flow (Split-And-Recombine, SAR) or recirculation patterns within the liquid plugs of segmented liquid/liquid flows. The main limitation to an increase of the mixing process following the above mentioned solutions is represented by the viscous dissipation.

Improvements are even possible by deforming the lamellae arrangement into tendril, whorl or striation-like shapes (Multilamination), which increases the interfacial area and the concentration gradient by reducing the striation thickness.

To describe the mixing process, an useful dimensionless number is the Fourier number, F_0, defined as the ratio between the average residence (Tr) and diffusive mixing time (Tm). Tr represents the ratio between the length of the channel L and the average velocity U; the diffusive mixing time (Tm) is the time that a molecule takes to diffuse a distance l:

$$Tr = \frac{L}{U} \tag{8.3}$$

$$Tm = \frac{l^2}{D} \tag{8.4}$$

Long Tr and short Tm favor the mixing of solutions; therefore high F0 values mean an efficient mixing. In general, the smaller the molecule, the larger is the diffusion coefficient D and the faster the molecule can diffuse. Beside the diffusion coefficient, a crucial parameter to evaluate the mixing time due to diffusion is the relevant length for mixing Lm which corresponds to:

$$Lm \sim U \times \frac{l^2}{D} \sim Pe \times l \tag{8.5}$$

For laminar uniaxial flows, the mixing length Lm was found to be proportional to Pe times the channel distance l.

The mixing performance is expressed by the mixing efficiency α at any cross-section of the microchannel (MC). It can be computed via the following formulation (Erickson and Li 2002):

$$\alpha = \left[1 - \frac{\int_{\text{lowersurface}}^{\text{uppersurface}} \left|C^* - C_\infty^*\right| dy^*}{\int_{\text{lowersurface}}^{\text{uppersurface}} \left|C_0^* - C_\infty^*\right| dy^*}\right] \times 100 \tag{8.6}$$

where C^* is the non-dimensional species concentration at any point across the width of the channel; ${C_\infty}^*$ and ${C_0}^*$ are the non-dimensional species concentrations at the corresponding point in the perfectly mixed and unmixed conditions. A fully mixed state would have a 100 % mixing efficiency, while the unmixed state would have a 0 % mixing efficiency. However, the mixer efficiency is influenced by the microreactors geometry and can be maximized by designing low pressure drop static mixers.

The pressure drop ΔP is due to frictional forces caused by the resistance of a fluid to move through a static mixer, and is defined as:

$$\Delta P = (4f)\left(\frac{L}{d}\right)\frac{\rho V^2}{2} \tag{8.7}$$

Where f is a non-dimensional friction factor coefficient, ρ the density of the fluid, V the superficial velocity, d the characteristic diameter of the static mixer, and L is the total length of the micromixer.

An inefficient mixer design will have a relatively high pressure drop and/or a long mixing length; on the contrary, an efficient mixer design will have a relatively low pressure drop and/or a short mixing length. With zero energy consumption (no pressure drop) and instantaneous mixing (zero length), the ideal mixer would have a theoretical 100 % mixing efficiency.

The product between the pressure drop and the volumetric flow rate (defined as the volume of fluid which passes through a given surface per unit time) is the energy dissipation (*Ed*) inside the micromixer, a key parameter in many operations like in emulsification or mixing.

Dimensionless numbers like *Re*, *Pe* and F_0, provide insight into the expected performance of a particular mixing design; however the actual performance maybe predicted only by full computational fluid dynamics (CFD) which have been discussed in Chap. 3.

Under an experimental point of view there are some approaches to characterize micromixing which are based on competing reaction and diffusion rates. The Villermaux-Dushman method, for example, was adapted to continuous processes, and improved in sensitivity to allow iodine UV–Vis detection in real-time at different flow rates. Methods for optimizing mixing in multiphase reactions at microscale based on dye extraction have been also developed (Panić et al. 2004).

8.2 Overview on Micro Mixing Technology

As discussed previously, an efficient mixing is the first step to perform a number of chemical processes such as reactions, extraction, crystallization, distillation, etc. Therefore the parameters previously mentioned and summarized in Table 8.1 represent the basic rules to fabricate microdevices able to transfer all the synthetic processes from macro to microscale.

Very complicated and smart designs have been projected and realized in materials like glass, silicon, polymers or metals with planar or 3-D arrangements and a complete overview could be found in reviews available in the literature (Lee et al. 2011; Mansur et al. 2008; Nguyen and Wu 2005; Suh and Kang 2010). In the following we wish to describe the main features and geometric structures of the passive micromixers, the most widespread microfluidic devices integrable in Lab-On-a-Chip systems used by the organic chemists interested in flow chemistry.

Basically, two main categories of micromixers exist: passive and active micromixers. Both of them have been used for performing organic synthesis as well as crystallization or extraction processes. In *passive mixers*, the flow energy is due to pumping action or hydrostatic potential; as a result, a faster mixing can be obtained (a) by restructuring of the flows creating thin lamellae, (b) by dividing and recombining the lamellae (SAR) (c) by inducing chaotic advection, (d) by forming droplets in mixed liquids (segmented flow), (e) by nozzle injection in flow or collision of jets and (f) by Coanda-effect (Hessel et al. 2005b). For several applications of flow chemistry (a–c) based-methods are preferred. Bi- and multilaminations are obtained in T or Y-flow and interdigital or bifurcation structures, while SAR with repeated flow division and recombination structures. Chaotic

Table 8.1 Basic rules to achieve a high efficient mixing in laminar flow regime (Re <2,300)

Theoretical parameter	To be minimized (↓) or maximized (↑)	Structural and operational parameters
D_f	↑	Large A via multilamination or SAR mechanisms
F_0	↑	Parameters to maximize Tr and minimize Tm
Tr	↑	Long channels, low fluid speed
Tm	↓	Small molecules with large D
Lm	↓	Molecules with large D, low fluid speed
ΔP	↓	Short channels, low fluid density, large hydraulic diameter
α	→ 100 %	Optimization of all the other parameters to achieve it in low pressure drop short microchannels
Ed	↓	Low pressure drop ΔP, small volumetric flow rate

advection is induced by introducing herringbone structures into a MC, by producing meandering or zigzag channels (Hessel et al. 2005b).

In *active micromixers*, external energy inputs are generated to increase mixing efficiency. Among the most important energy sources there are: ultrasound/acoustic/piezoelectric vibrating membranes, bubble-induced vibrations, electrokinetic instabilities or electrowetting-induced merging of droplets, periodic variations of the flow rates or integrated microvalves/pumps (pressure field disturbance), magneto-hydrodynamic action, small impellers, or thermal disturbance (Hessel et al. 2005b).

Active micromixers are generally more complex as it appears more difficult to operate, fabricate, clean, and integrate them into microfluidic systems. Therefore passive mixers are more commonly used in most microfluidic applications and this will be discussed in the following sections.

T/Y- mixers (Fig. 8.1a, b) have been used for several reactions and crystallizations (Ståhl et al. 2001; Haselhuhn and Kind 2003; Schwarzer and Peukert 2004; Schwarzer et al. 2006; Iwasaki and Yoshida 2005; Nagaki et al. 2008); serpentine, zigzag channels (Fig. 8.1c, d) for testing a variety of chemical reactions (Chambers and Spink 1999; Wiles and Watts 2011); multilamellae systems (Fig. 8.1e) for reactions, extraction, emulsification (Sprogies et al. 2008; Mitchell et al. 2001; Suga et al. 2003); SAR mixers (Fig. 8.1f) for organic synthesis (Hessel et al. 2005a), extraction (Mae et al. 2004; Okubo et al. 2004) and crystallization (Jung et al. 2012). Chaotic (Fig. 8.1g) and packed beds microreactors were applied to several catalytic reactions (Moon et al. 2010; Ziegenbalg et al. 2010; Rebrov et al. 2012; van Herk et al. 2009). Packed microcolumns have been often used for liquid chromatography at microscale (Kutter 2012).

The following paragraphs are addressed to microreactor designers that wish to fabricate chips of suitable geometries for a specific application. The structural constraints of the microreactors of current use in organic chemistry are outlined;

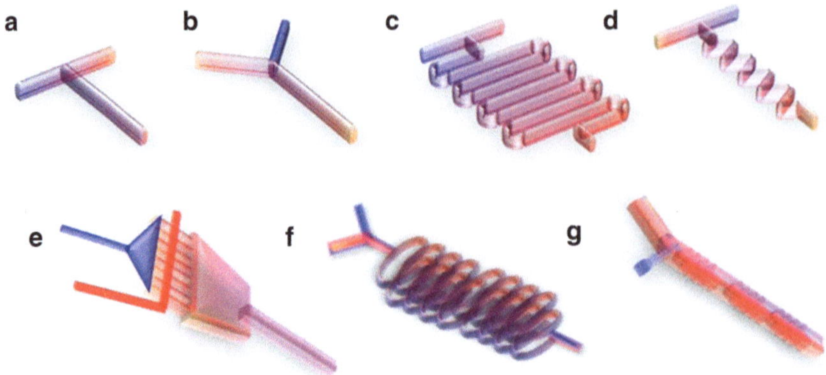

Fig. 8.1 Examples of (**a**) T-mixer, (**b**) Y-mixer, (**c**) serpentine, (**d**) zigzag, (**e**) multilamination, (**f**) SAR, and (**g**) chaotic flow micromixers

parameters like mixing efficiency, mixing length, pressure drop in the range of different *Re* or *Pe* numbers are described in detail.

8.2.1 T-Mixer and Y-Mixer

Liquid mixing was studied in T/Y-mixers (see Fig. 8.1 a and b) by A. Soleymani et al. (Soleymani et al. 2008) who elaborated a model on the basis of a simple system consisting of a T-channel in which they simulated the movement of a flow liquid at different flow rates. The dependence of the mixing performance α was related to three parameters: the *flow rates*, the *pressure drop* and the *depth/width* (*d/w*) aspect ratio.

In T/Y- mixers, the mixing performance decreases at higher *flow rates* (shorter residence times) at low *Re* (12), when the flow is strictly laminar and the mixing is entirely due to molecular diffusion between the layers of different concentrations (stratified flow). At moderate *Re* (80), a α increase indicates that convective mass transfer promotes mixing but the main mixing mechanism is still diffusion (vortex flow). At higher *Re* (240), complex convection mechanisms induce vortex generation and reduce mixing distance (engulfment flow).

At low *Re* the *pressure drop* profile is linear along the channel length; at moderate and high *Re* it exhibits a sharp decrease in the region of formation of vortices due to convection. Pressure drop was found to be dependent on the mixing angle variation and therefore to influence the mixing efficiency. In liquid systems inlet angles of 90° and 105° corresponded to the geometries for which vorticity reaches a maximum; no effect of injection angle was observed for mixing in gas phase (Gobby et al. 2001).

At very low *d*/*w aspect ratio* and constant flow rate, the high wall friction of the fluid generates a stratified flow regime and high values of pressure drop. The mixing is very low being dominated by diffusion. At intermediate aspect ratios, the pressure drop decreases and a transition from stratified to vortex flow occurs. In a certain *d*/*w* range vortices' formation is maximized as well as the mixing efficiency. After reaching the optimum, an increase in the *d*/*w* ratio produces a damping of the vortices, and decreases α. A similar behavior was also observed in the case of mixing in gas phase (Gobby et al. 2001).

No significant mixing efficiency improvement was observed with the introduction of a throttle in a T-mixer in the case of liquid mixing; on the contrary, throttling gas flows strongly decreases the mixing lengths (Gobby et al. 2001).

In conclusion, in order to maximize α of liquid phase processes, efficient T/Y MCs of rectangular cross-section have to be designed following the guidelines:

- Inlet angles of 90° or 105°.
- Channel *d*/*w* aspect ratio between 0.5 and 0.7.

8.2.2 Zigzag and Curved Micromixers

The creation of recirculation flow patterns is reported to increase the mixing efficiency at sufficiently high *Re* (Mengeaud et al. 2002). The addition of zigzag channels to T or Y mixers (see Fig. 8.1d) produces such an effect and improves the mixing process. In a first pioneering study (Mengeaud et al. 2002) the flow circulation dynamics was studied in channels with a number of *s*/*w* ratios (see Fig. 8.2, *s* is the linear length of the periodic zigzag step and *w* is the width of the channel) at constant *Pe* numbers and *Re* ranging from 1 to 800. As reference, three linear Y-mixers (like those shown in Fig. 8.1b) of different lengths and cross sections were chosen. A reduction of the cross section or the transversal diffusion length of 29 % was demonstrated to enhance the mixing efficiency of 95 %; on the contrary, a 29 % length increase generated an enhancement of the mixing efficiency of only 13 %. Therefore it was clear that *the cross section influences the mixing efficiency more than the mixing length.*

At low *Re*, the mixing efficiency did not show any improvements with the addition of zigzag turns of *s*/*w* ratio ranging from 1 to 8.

An interesting effect was observed for the same geometries at $Re = 265$. A significant improvement in comparison with the reference linear mixer was observed at $s/w > 2$ with a parabolic profile achieving its maximum at $s/w = 4$ and then tending asymptotically to reference linear mixer value.

The mixing efficiency was also studied at fixed *s* as a function of *Re*. In such a case, a parabolic profile was observed with constant value of 81 % up to $Re = 80$ and yields reaching 98.8 % at $Re = 268$.

The finite element model elaborated by Mengeaud et al. (2002) finally demonstrated that *at high Re*, inertial effects induce a convergence of the flow lines along

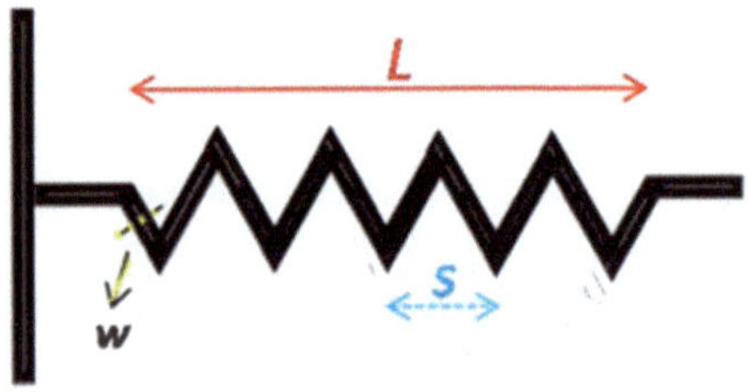

Fig. 8.2 Design and parameters of a zigzag micromixer (Mengeaud et al. 2002)

the zigzag walls and an increase of the local velocity, that *maximize the mixing efficiency at s/w ratios of 4*.

Besides zigzag patterns, micromixers can be also modified with *square-wave or curved MCs* (see Fig. 8.1 c). Hossain et al. (2009) compared the flow and mixing mechanisms of different channel geometries (zigzag, square-wave, and curved) at fixed height, axial length, channel width, periodicity and number of repeating units. A comparison of mixing quality of water and ethanol along the channel length at low *Re* revealed that for all the three geometries the mixing is dominated by the residence times and depends on the total path of the flow; consequently no differences between the geometries was envisaged. At *Re* of 20, the mixing capabilities of square-wave MCs were found to be 1.34 times larger than zigzag and curved MCs. At higher *Re* different behaviors were depicted. While at $Re = 50$ the mixing efficiency increased for all the three geometries, the opposite occurred at *Re* of 267. However, in square-wave MCs, the maximum mixing efficiency was achieved in the middle of the channel, for the other two geometries the entire length of the channel was needed. The faster mixing at higher *Re* in square-wave channels was explained in terms of a high degree of turning that accelerated the flow near the inner corner and decelerated at the outer corner flow during turning. The high degree of acceleration induced vortices.

Another important aspect that was evaluated was the *pressure drop* which increased rapidly with *Re* in all the MCs and was roughly the same at $Re = 50$ in square-wave and zigzag MCs. In curved MCs a smaller pressure drop was observed because of the loss of sharp turns in such geometry.

A recent modification of curved MCs consists of converging–diverging meandering structures with semi-elliptical side walls which increase fluid mixing via multidirectional vortices (Wu and Tsai 2013).

Therefore, zigzag, square-wave, or curved MCs allow:

- To increase the mixing efficiency at high *Re* compared with linear MCs.
- To maximize the mixing efficiency at high *Re* optimizing the *s/w* ratio of zigzag MCs around 4.
- A faster mixing in square-wave MCs (better than curved and zigzag patterns).
- A lower pressure drop in curved MCs than in zigzag or square wave patterns.

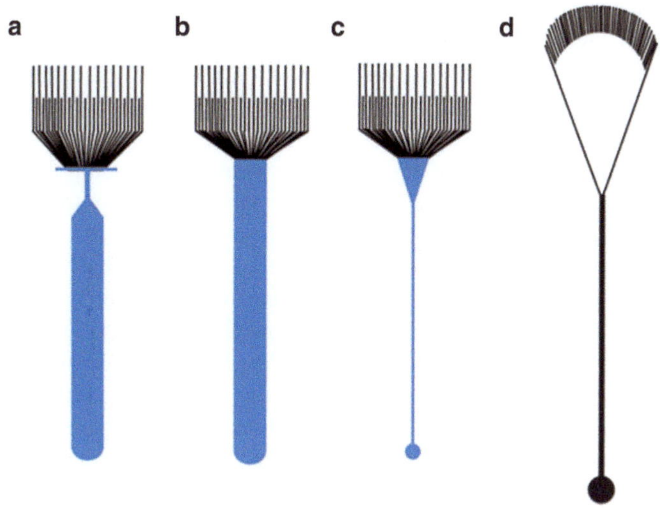

Fig. 8.3 Multilamination micromixer designs (**a**) slit-shaped, (**b**) rectangular, (**c**) triangular, and (**d**) Super Focus (Hardt and Schönfeld 2003)

8.2.3 Multilamination

For increasing the mixing efficiency of chemical reactions, an important strategy is to create as thin as possible fluid lamellae that diffuse faster than large lamellae in the microreactors dominated by laminar regime. *Interdigital channels* with parallel or counter oriented flows are the most common examples of multilamination microreactors (see Fig. 8.3 and Fig. 8.1 e).

The mixing is naturally faster than in laminar T-configuration flow and decreased mixing time can be achieved by generating thinner lamellae in very narrow channels, whose minimal dimensions are restricted by the capabilities of the manufacturing and by pressure loss limits (Hardt and Schönfeld 2003). In multilamination MCs, mixing time was calculated to be directly proportional to lamellae width (w_l) according to the law: w_l^2/D, being D the diffusion coefficient.

Interdigital microreactors consist of an array of multichannels focused into chambers of rectangular, triangular, slit-shaped or super focus geometry to allow fast mixing of the thin lamellae feeds (see Fig. 8.3). The focusing of fluid streams that is, narrowing of the lamellae width is induced with *hydrodynamic or geometric restriction.* In the hydrodynamic focusing, an increasing of the ratio between the two fluids reduces the layer width of one fluid at the expense of the other. A narrowing of the lamellae also results by the addition of geometric restrictions which may cause lamellae tilting (Hessel et al. 2003).

In rectangular, triangular, slit-shaped mixers with 2×15 feed MCs of 60 μm width (see Fig. 8.3), very different lamellae patterns, mixing efficiency, *Re* and *Pe* numbers at several flow rates have been calculated. Very regular lamellae patterns were observed in rectangular micromixers with lamellae width between 110 and

30 μm, *Re* 2–341 and *Pe* 6.41×10^3–1.07×10^6 at speeds of 12–200 mL/h. Unfortunately, the mixing was not completed even at low flow rates. In triangular micromixers, lamellae exhibit a curved shape with a non-homogeneous profile. Lower *Re* and *Pe* numbers than rectangular ones were calculated with a mixing efficiency higher at lower flow rates due to the longer residence time. The geometric focusing allowed reducing the lamellae width by a factor of 6. Slit-type microreactors were the most efficient with *Re* and *Pe* up to 1,702.94 and 5.34×10^6, respectively (see Fig. 8.3a). The flow pattern of the lamellae in the slit was not uniform and in the rectangular area a jet was formed at medium and high flow rates. The mixing time was in the range of milliseconds and was due to convective phenomena, typical for turbulent mixing. In a special design of the triangular mixer, named *Super Focus* (see Fig. 8.3d), liquid mixing times are reduced to about 10 ms and lamellae are regular and compressed by a factor of 40, from 160 to 4 μm. The Super Focus was realized with precise design constraints to respect pressure loss and fabrication issues. The result was a system with 124 MCs of 100 μm width focused through a triangular funnel into a long 500 μm wide (parameter *b*) rectangular MC. The *a*/*b* *ratio* (*a* represents the length of the triangular channel) was maximized (about 40, considering the fabrication limitation to fit a microscope slide) and the length of the rectangular channel was chosen to assure complete mixing even at high flow rates (1,000 mL/h). At 8,000 mL/h (maximum flow rate) the *Re* corresponded to 4,427 and *Pe* of 1.39×10^7.

8.2.4 Splitting and Recombination (SAR)

Split-and-Recombine mixers create multi-laminar flow patterns, but with a strategy different from interdigital feed systems. Three essential steps are required: (1) flow splitting, (2) flow recombination and (3) flow arrangement or reshaping (see Fig. 8.4). Several geometries able to act as SAR mixers have been designed embodying forks, ramps, curves or splitting planes. The main effect of these architectures was to generate a secondary recirculation flow pattern at high *Re* for most liquid and at $Re < 15$ for highly viscous liquids. Chaotic advection, which generally induces regions of regular flows and poor mixing besides regions of high chaos at high mixing occurrence, was demonstrated to be responsible for the high spatial homogenous mixing observed in SAR mixers (Schonfeld et al. 2004). It is well known that mixing efficiency α depends on diffusive mass transport which is related to the interfacial area. In SAR mixers the interface area is proportional to the repetition of the three steps for a number n of elements and is dependent on the mean velocity u and the length l_{SAR} of one SAR unit:

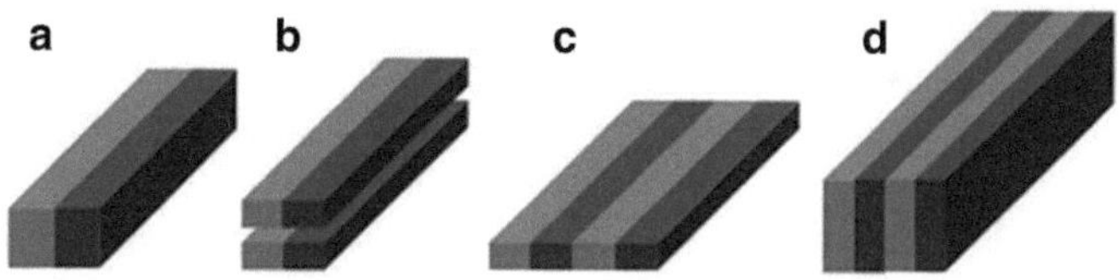

Fig. 8.4 Schematic of a SAR approach: (**a**) initial configuration, (**b**) splitting, (**c**) recombination, (**d**) reshaping (Schonfeld et al. 2004)

$$\alpha \propto 2_{l_{\mathrm{SAR}}}^{ut} \tag{8.8}$$

A SAR mixer widely used in the synthesis of organic compounds or emulsions was the Caterpillar mixer from IMM (Institut fur Mikrotechnik Mainz GmbH, Mainz, Germany, http://www.immmainz.de/index.php?id=caterpillar_micromixer) which was used to perform aqueous Kolbe-Schmitt synthesis using resorcinol under high pressure and temperature conditions (Hessel et al. 2005a) or crystallization of polyethylene submicron particles (Jung et al. 2012). In such a system, starting with a two lamellae configuration of 1,200 × 1,200 mm^2 cross section, after eight mixing steps, the fluid is ideally divided into 512 lamellae of 2.4 mm width. The pressure drop of viscoelastic fluids begins at very low *Re* due to the high elongation flow present in Caterpillar static mixers.

Caterpillar mixers are manufactured in steel, but similar devices have been realized in glass or silicon even with smaller channels. Curved geometries of each elements oriented perpendicularly to the flow direction have been designed and organic reactions or emulsions, crystallizations and extraction processes performed inside (Zuidhof et al. 2012; Okubo et al. 2004). In the design of each mixing element, several parameters to improve mixing efficiency have to be taken into consideration. In the realization of σ-type plate static mixers (see Fig. 8.1f), for example, two polytetrafluoroethylene (PTFE) plates containing the structures were sandwiched between two acrylic resin plates and a Y-channel was used for liquid injection into the first element that allows the fluid placement parallel to each other at the exit. The fluids merge in a two layer stream in the interconnecting circular channel and then enter in the second element where they are split into two sub-streams (Ohkawa et al. 2008). The process of merging and splitting is repeated into the second element with a 180° turning of the substreams and the generation of four substreams which fed the third element. The process is repeated and a multilamellae stream is generated with mixing improvement.

However, the predicted fluid circulation mechanism appears to be strongly influenced by the *d*/*w* *aspect ratio* of the σ-type elements and the number of elements. It was observed that in shallow channels (low ratio) the fluids flow with incomplete recombination and insufficient multilamination. At higher ratios (1 to 5), parallel flow streams are produced and a recombination closer to the ideal condition obtained.

A deviation from the ideal mixing behavior is also observed at $Re > 10$, since the secondary flow, generated around the two-and three-dimensionally curved portions of the mixer, distorts the progress of systematic splitting and recombination. A dependence of the *number of elements* needed to achieve complete mixing and the *Re* was found. It was demonstrated that for two liquids with almost the same physical properties at $Re < 10$, n increases with *Re* or flow rate because of a shorter residence time. At $Re = 10–20$, n becomes maximum; afterwards it decreases a lot and becomes less than 10 at $Re > 10^2$ due to the elongation and deformation of the fluid interface caused by the secondary flows. The *pressure drop*, directly proportional to the friction factor, was larger than that for straight square channels. At $Re < 10$, ΔP was found to be proportional to Re^{-1} and therefore its variation decreases with further *Re* increasing.

8.2.5 Chaotic Flow Micromixers

The introduction of microstructures inside MCs causes flow circulations which lead to an exponential increase of the specific interface, and generates chaotic advection, thus improving the mixing. Herringbone (Stroock et al. 2002), cylindrical (Wang et al. 2012) and tooth (Kim et al. 2005) type structures have been proposed and integrated into one or two channels, with symmetric or asymmetric design and with the addition of barriers (Kim et al. 2004). The most representative example of this group of microreactors is the staggered herringbone mixer (SHM) that was used for a number of catalytic applications (Moon et al. 2010; Ziegenbalg et al. 2010; Rebrov et al. 2012). The device consists of a MC with several of sets of herringbones alternated with complementary ones (see Fig. 8.5).

Each set of alternated grooves cause transverse flow in the channel, resulting in two counter-rotating vortices along the channel length. The alternation of counter-rotating vortices reorients the flow periodically, disrupting the relatively untouched elliptic points in each of the vortices and dramatically improving mixing. The effect of individual half-cycles is a non-chaotic process, but the reorientation of the circulating flow produces chaotic mixing. The effect is evident even at low *Re* where, in simple MCs, pressure flows are laminar and mixing purely diffusive.

Under a practical point of view, it was calculated that a stream of protein solution in aqueous buffer which flows at a speed of 1 cm/s in a MC with a cross-section of 0.01 cm ($Pe = 104$) has a mixing length in a simple MC of 100 cm.

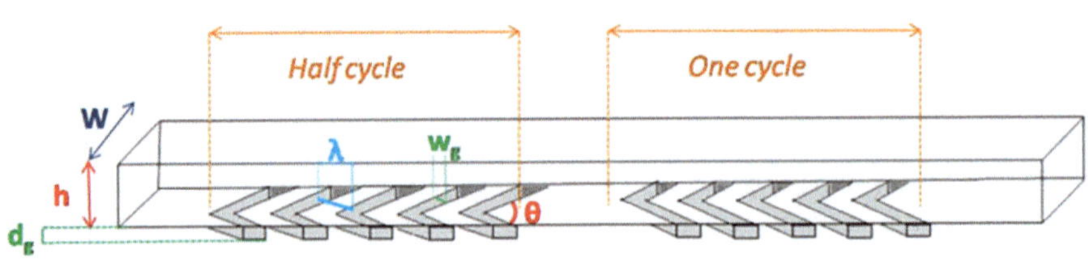

Fig. 8.5 Staggered herringbone mixer (SHM). (**a**) Schematic diagram of one-and-a-half cycles and main geometric parameters (Stroock et al. 2002)

In a SHM with the same cross-section, the mixing length corresponds to 1 cm. Furthermore, increasing the flow speed by a factor of 10 (i.e., to $Pe = 105$) the mixing length in the SHM increases to 1.5 cm. With the same change in flow speed, the mixing length in the absence of stirring increases tenfold, to 103 cm. These results are due to the dependence of the mixing length on $ln(Pe)$ rather than Pe for chaotic flows at higher speed (Stroock et al. 2002).

Confocal studies of cross section of the flows allowed to identify two basic requirements to observe chaotic mixing in MCs in presence of grooves at most of the cross-sectional areas: p, a measure of the asymmetry of the herringbones that is ½ for symmetric herringbones, must be around 2/3 and $\Delta\varphi_m$, a measure of the amplitude of the rotation of the fluid in each half cycle which goes to zero for non-chaotic flows, has to be >60° (Stroock et al. 2002).

However, under a practical point of view, to optimize the design of a SHM and therefore achieve a high mixing performance with the lowest pressure drop possible, some structural parameters have been demonstrated to play an important role (Cortes-Quiroz et al. 2010). The *ratio between the groove width* w_g *and groove pitch* in direction perpendicular to the walls of the grooves λ (w_g/λ), was calculated to give the highest mixing performances at a value of 0.75, taken into consideration limitations due to materials and fabrication issues (see Fig. 8.5). An increase of the mixing index and a decrease of the pressure drop were also observed for increasing *the ratio between the depth of the groove* d_g *and height of the channel* h (d_g/h). An optimum value of 0.60 was calculated after optimization with fixed aspect ratios. The effect of the parameter θ (see Fig. 8.5) on the mixing was not so explicit as the others; when θ increases from 70° to 80° it was observed an increase of pressure drop and a slight increase of mixing performance. Additionally, no improvement was reported with an increase of the number of elements from 5 to 7 (Cortes-Quiroz et al. 2010).

In conclusion, there are several tools to improve mixing efficiency based on the interdiffusion of thinner lamellae or on the introduction of vortices and chaotic flow regimes. Both the mechanisms are obtained by designing specific geometries of mixers and minimizing the internal pressure drop to reduce energy dissipation. A summary of the above described mixers, the most important structural and operational parameters is reported in Table 8.2.

8.2.6 Packed Beds for Catalytic Reactions

Packed beds microreactors have been used to perform several catalytic reactions and for liquid phase chromatography. The geometry of the devices and the powder used to fill them are different according to the function they are designed for.

For catalytic two or three phase reactions, serpentine circular microreactors are filled with catalyst and variable percentage of diluent powders.

Table 8.2 Structural and operational parameters to maximize mixing efficiency in several micromixers

Micromixer	Structural parameters	Effects	Operation conditions	Reference
T/Y- mixers	inlet angles of 90° or 105°; *d*/*w* about 0.5–0.7; long channel length	↑α	Re >120, high speed flow	(Soleymani et al. 2008)
Zigzag, square, and curved mixers	*s*/*w* zigzag turns ≈4	↑α	Re >80	Mengeaud et al. 2002; Hossain et al. 2009
	Curved < Square ≈ Zigzag turns	↓ ΔP		
	Square < Zigzag < Curved turns	↓Tm		
Multilamination	Thin Interdigital microchannels	↓Tm	High flow rates (up to 1,000 mL/h)	Hessel et al. 2003
	Super Focus >> slit-shaped > triangular > rectangular design	↓ΔP, ↓Tm		
SAR	High number *n* of repetition units; d/w aspect ratio of the elements ≥1	↑α	Re < 10, *n* proportional to Re and ΔP to Re^{-1}; Re > 100, *n* < 10	Ohkawa et al. 2008
Chaotic flow	$w_g/\lambda = 0.75$	↓ Lm	High speed	Cortes-Quiroz et al. 2010
	d_g/h about 0.6	↑α, ↓ΔP		

Table 8.3 Features of some important radionuclides and their use in nuclear medicine

Nuclide	Half-life	Imaging apparatus	Examples
^{99m}Tc	6 h	SPECT	^{99m}Tc-Sestamibi (myocardial perfusion), ^{99m}Tc-MAG3 (renal function), ^{99m}Tc-HMPAO (cerebral perfusion)
^{123}I	13.2 h	SPECT	^{123}I-NaI (thyroid function), ^{123}I-MIBG (cardiac function)
^{18}F	110 min	PET	^{18}F-FDG (glucose metabolism), ^{18}F-FDOPA (dopaminergic function), ^{18}F-FLT (tumor imaging)
^{11}C	20 min	PET	^{11}C-raclopride (D2 receptor imaging), ^{11}C-methionine (tumor imaging), ^{11}C-choline (prostate cancer imaging)
^{68}Ga	68 min	PET	^{68}Ga-DOTATOC (Somatostatin expressing tumors)

Table 8.4 Relevant milestones for utilization of microfluidics in radiochemistry

Tracer/reaction	Modality	Notes	Reference
^{11}C-methylations/^{18}F-fluoroalkylations	MCS	1st report of use of microfluidics for radiochemistry	(Lu et al. 2004)
^{18}F-FDG	MVS	1st report on integrated microfluidic system for radiochemistry	(Lee et al. 2005)
^{18}F-CB102	MCS	Proof of principle for DOD	(Pascali et al. 2010)
^{18}F-fluoroalkyl cholines	MCS	1st report on successful 2-step/1-flow process	(Pascali et al. 2011)
^{18}F-fallypride	MVS	1st report on a microfluidically produced tracer used for human injection	(Lu et al. 2009)
^{64}Cu-DOTA-cyclo (RGDfk)	MVS	PDMS microreactor for radiometals	(Wheeler et al. 2010)
^{18}F-phenylmethylazides	MCS	Microfluidically enhanced radiochemistry studies	(Chun and Pike 2012)
^{11}C-carbonylations	MCS	Microfluidically enhanced radiochemistry studies with ^{11}C	(Miller et al. 2011)
^{13}N-azacompounds	MCS	1st report on use of ^{13}N in microfluidics	(Gaja et al. 2012)
^{18}F -FB-MQSpTPL-sC18	MCS	1st report on microfluidic optimization of protein labeling	(Richter et al. 2012)
^{18}F preconcentration	Micro-SPE	Custom-built micronized system	(De Leonardis et al. 2011)
^{18}F electrochemical preconcentration	MCS	Custom-built micronized system	(Sadeghi et al. 2013)
^{18}F-FDG purification	Micro-SPE	Custom-built micronized system	(Tarn et al. 2013)
Several ^{18}F tracers	MVS	EWOD based prototype	(Keng et al. 2012)
Several ^{18}F tracers	MCS	Integrated system with electrochemical ^{18}F preconcentration	(Wong et al. 2012)
Several ^{18}F tracers	MCS	Interfacing microfluidics with additional hardware	(Pascali et al.)

The *bed length* L_b of the microreactor (usually in the range of few centimeters) is a parameter to be taken into consideration, to ensure axial dispersion and therefore the complete wetting of the catalyst:

$$L_b > 20 n_K \frac{d_p}{Pe_p} \ln(1 - X) \tag{8.9}$$

Where X is the conversion, n_K is the reaction order, d_p is the particle diameter, and Pe_p is the particle Peclet number. The addition of a small quantity of diluents to the catalysts and the uniform homogenous package of the powders in the bed are important to obtain the calculated bed length. A crucial point is, therefore, to optimize the catalyst-loading procedure and the catalyst/diluents ratio to get a bed without any segregation. The presence of particle segregation were demonstrated to cause preferential flow dynamics, particularly when liquid and gas phases are fluxed to react (van Herk et al. 2009). Usually, voids between the smaller particles remain liquid-filled and gas does not penetrate these pockets. The main effect is poor yields and low reproducibility. In some catalytic applications, diluent and catalyst powders of similar free-fall velocity were used to minimize segregation. The dependence of the conversion yield on the *dilution ratio* was also investigated coming to the conclusion that a small amount of diluent (volumetric diluents/catalyst ratio less than 5) does not reduce the conversion; on the contrary, dilution ratios higher than 10 induce a drastic conversion decrease (van Herk et al. 2009). The control over parameters like diluents/catalyst ratio and dimensions was reported to be more determinant than the choice of operation liquid velocities below 0.1 mm/s for diluents size smaller than 200 μm (Bej et al. 2000, 2001) that was considered an important requirement before van Herk's experiments. However, the introduction of solids in MCs limits fluid flow, influence mass and energy transports, and render continuous reactors inoperable (Hartman 2012).

To reduce pressure loss and allow for a better flow of reagents, catalyst can be deposited on the walls of microreactors using chemical methods like sol–gel, chemical vapor deposition (CVD), direct synthesis or physical methods like physical vapor deposition (PVD) (Meille 2006).

As mentioned above, packed beds are also used for liquid phase chromatography (Kutter 2012). For such application it is mandatory to design MCs of suitable *w/d aspect ratio* to avoid the "wall effect" due to differences in the quality or density of the packing or the flow behavior in the bulk or close to the wall of the chips. The "wall effect" is more pronounced for the typically used ratios in microfluidic systems (around 10, i.e., channel width or depth is about 10 times the used particle diameter or equivalent feature size). For small values (particle size of the same order as channel depth) or large values (very small particles or relatively large channels) the effects are much less pronounced, either because the wall and the bulk region become very similar or because the wall region only constitutes a small part of the entire channel cross section. Another geometric constraint is the channel *cross-section shape* to avoid corner region of more stagnant flow. Additionally, if the electro-driven flow is used to move the mobile phase, the impact of joule heating on the separation performance has to be considered.

The stationary phases packed as beds can be (a) particles modified with the retentive material or self-assembled inside the channel; (b) monoliths, i.e., a porous scaffold, which either already includes the retentive species or can be grafted with it later; and (c) micromachined pillar arrays, whose surfaces can be modified to increase retention.

To determine the efficiency of microcolumns the most important *figures of merit* are the number of *theoretical plates (N)* and the height equivalent to a theoretical plate (H) to determine how fast the system can be driven and hence how fast a separation can be delivered. Moreover, the *linear velocity of the mobile phase (v)* has to be modulated. Plots that relate H and v are the well-known van Deemter plots, ($H = A + B/v + Cv$) a closer analysis of which can yield insight into the contribution of different mechanisms to H, namely the contribution from flow path differences (eddy diffusion, A term), the contribution from longitudinal (or axial) diffusion (B term, equivalent to the Péclet number formalism), and the contribution from mass transfer kinetics (C term). Van Deemter plots often show a minimum, which corresponds to the best performance and the linear flow rate at which it can be achieved. Driving a system faster or slower typically results in a decreased performance. The A-term contribution can be reduced using carefully designed microfabricated pillar arrays or "exquisitely" packed particulate beds. The B-term contributions are minimized at sufficiently large linear velocities of the mobile phase (Kutter 2012). Because of the smaller dimensions involved in miniaturized separation systems, and the corresponding reduced diffusion distances, the C term often has a very shallow slope, thus allowing operation at higher velocities without sacrificing a lot of separation efficiency.

However, van Deemter equation does not take into account the increase in pressure drop with the square of the particle size which limits the maximum column length or mobile-phase velocity that can be employed; this affects the maximum plate number that can be generated or the minimum required analysis time. A prediction of these aspects not explicitly covered by the van Deemter formalism is possible by investigating the kinetic plots (Causon et al. 2011).

8.3 Use of Microfluidics for Chemical Synthesis

For centuries, organic chemists have almost exclusively used batch reactors (test tubes, round bottomed flasks etc.) to perform all chemical reactions and manufacturing processes. Although automated technology such as combinatorial synthesizers, are used in modern industry, fundamentally the reaction is still conducted in batch mode. However from a process chemistry perspective, a major problem observed with conventional batch technology is the failure to scale-up successful reactions; this is particularly problematic for highly exothermic processes. In industry this often means reoptimization of synthetic routes on several occasions when transferring from laboratory to pilot scale and then finally to production.

More recently the synthetic chemistry community has started to investigate the use of microfluidics as a tool for organic synthesis; however it would be fair to say that this has been a slower transition than that seen in the micro analytical arena. As already discussed in the first part of this chapter, microfluidic systems exhibit excellent mixing and heat transfer characteristics; consequently a key advantage of flow reactor technology for the organic chemist is the ability to very accurately control reaction parameters. For instance, the regulation of temperature and concentration is crucial in maintaining control over a synthetic reaction, not only to ensure selective product formation, but also from a safety perspective. Due to the excellent heat and mass transfer, and predictable flow properties exhibited by microfluidic reactors a high degree of reaction control is attainable. Along with increasing the rate of mixing, decreasing the channel diameter of the microfluidic reactor results in an inherently high *surface/volume ratio* (SVR), allowing rapid dissipation of any heat generated over the course of a reaction. Specific examples of up scaling are given at the end of this chapter.

In the following, selected examples are used to illustrate the advantages of the microreactor technology; more detailed information is cited in review articles (Wiles and Watts 2011, 2012) and books.

8.3.1 Microfluidic Reactors

Microfluidic systems or just microreactors are most commonly regarded as planar devices having channel dimensions in the range of a few hundred microns and may be fabricated in a number of materials such as silicon, quartz, glass, metals, and polymers. They may just contain a simple T- or Y-mixer, or may be fabricated to include the more complex mixing geometries as addressed earlier in this chapter. These reactors generally have volumes ranging from nanolitres to microliters; consequently these systems are most commonly used in small scale synthesis and screening processes, as well as niche areas such as radiochemistry discussed later in this chapter.

One important consideration for synthetic applications is chemical compatibility of reactor material with the reagents being used; consequently glass reactors are the most widely reported. The second most common material for reactors of this type is silicon. From a synthesis perspective, another advantage of microfluidic reactors is the very efficient thermal transfer, with heat transfer coefficients of the order of 60,000 $W/m^2/K$ compared with 100's $W/m^2/K$ for batch vessels (Morini 2004). Consequently reaction exotherms can be readily dissipated, enabling highly energetic materials to be synthesized in a safe manner.

The very small reactor size means that the "dead volume" of the system is very low and hence very short equilibration times are needed; with the consequence that very rapid reaction optimization is possible. As an example, Wiles (Wiles and

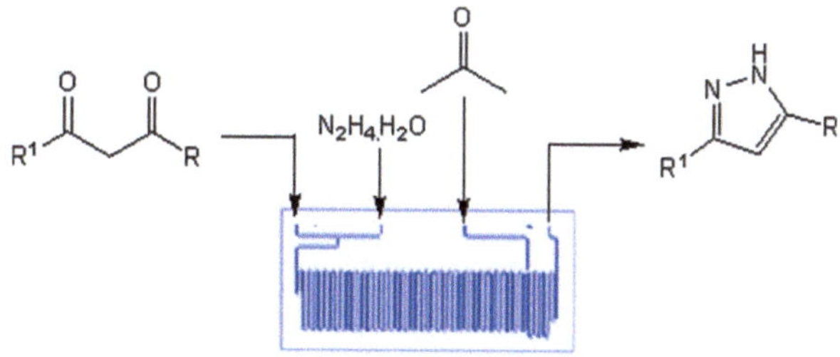

Fig. 8.6 Microreactor design used in azole synthesis

Watts 2012) performed a study to investigate the reaction between diketones and hydrazine to prepare a range of azoles (Fig. 8.6) using a 10 μL glass reactor inserted within an automated microreactor platform.

In these microfluidic reactors, processes are optimized by sequentially conducting a series of reactions at different residence times and temperatures in order to optimize the reaction conditions. In the above example, the reaction was evaluated at eight temperatures (25–195 °C) and five residence times (30 s to 5 min) using a variety of solvents. In this process a total of 200 experiments were conducted in 27 h using less than 6 mL of solvent. The optimum conditions were found to be a reaction temperature of 125 °C and residence time of 180 s using ethanol as the reaction solvent (Fig. 8.7).

As a more sophisticated extension, McMullen and coworkers (2010) developed an integrated process for the "self-optimization" of reactions performed within a microfluidic reactor. The authors demonstrated the ability to integrate feedback control into the reaction optimization process, enabling a dramatic increase in speed compared to conventional design of experiment (DoE) approaches. Selecting the Heck reaction of 4-chlorobenzotrifluoride and 2,3-dihydrofuran as a reaction (Fig. 8.8), they identified $Pd(OAc)_2$, *tert*-butyl-MePhos as the ligand and *n*-BuOH as the best reagents and solvent mixture, to form the product; reporting that the optimal conversion of 88 % was obtained using a residence time of 10 min at 90 °C.

Utilizing a capillary based microreactor interfaced to ultrahigh performance liquid chromatography system, Fang et al. (2010) demonstrated the development of a high-throughput screening system for homogeneous catalysts, focussing on the evaluation of catalyst efficacy towards the Friedel–Crafts cyclization (Fig. 8.9). Connecting the capillary reactor to the auto injector of the chromatography system, they were able to introduce the catalyst and reactant solution into a continuous carrier stream affording small, discrete reaction zones within the capillary. Using this approach, the authors evaluated the effect of a series of Lewis and Bronsted acids along with reaction temperature (0–40 °C) and molar ratio. The authors were able to demonstrate the screening power of their technique whereby 18 catalysts were assessed in 6 h; which enabled the identification of $Er(OTf)_3$ as the most efficient catalyst. In addition to providing a rapid technique for the screening of catalysts using low volumes of reactants (*ca.* 400 μg for each reaction), the methodology proved to be reproducible.

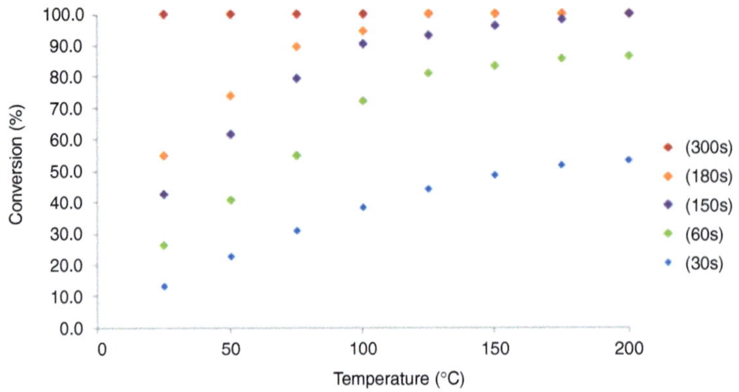

Fig. 8.7 Reaction optimization of azole reaction

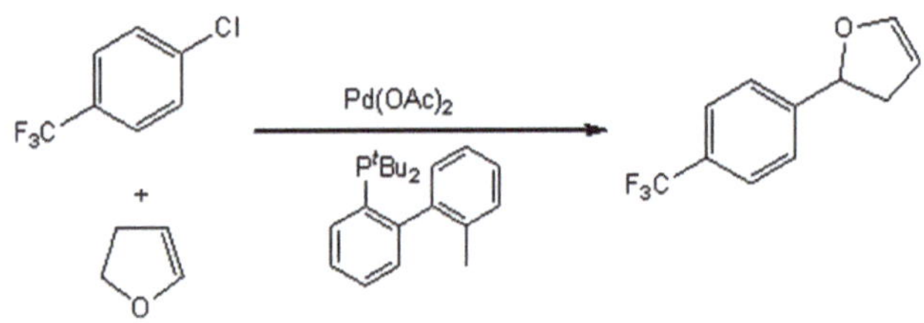

Fig. 8.8 Heck reaction illustrating catalysis in a microreactor

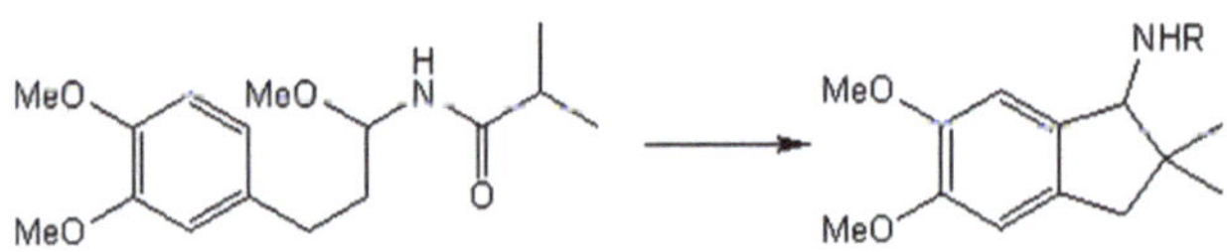

Fig. 8.9 Reaction used to illustrate high throughput screening utilizing a microreactor

As outlined, microreactors have attracted substantial interest with regard to improving safety. A particularly problematic area of synthetic chemistry is that of diazonium chemistry, where hazards include light, heat, and shock sensitivity, which can all lead to uncontrollable decomposition of the diazonium salts formed with the risk of explosion. These hazards are amplified when isoamyl nitrite is used in the preparation of the diazonium salts, as it is prone to decomposition even when refrigerated, limiting its use on an industrial scale. Fortt et al. (2003) addressed these issues by investigating the synthesis of chloroarenes in a glass microreactor (channel dimensions, 150 μm (wide) × 50 μm (deep) × 3.6 cm (length)). The first step of the reaction involved treatment of the amine, with isoamyl nitrite, at room temperature, to afford the diazonium salt (Fig. 8.10), followed by halogen abstraction from copper chloride, at 65 °C, to afford the respective chloroarene. Conducting the micro reaction in DMF as solvent 71 % conversion was reported,

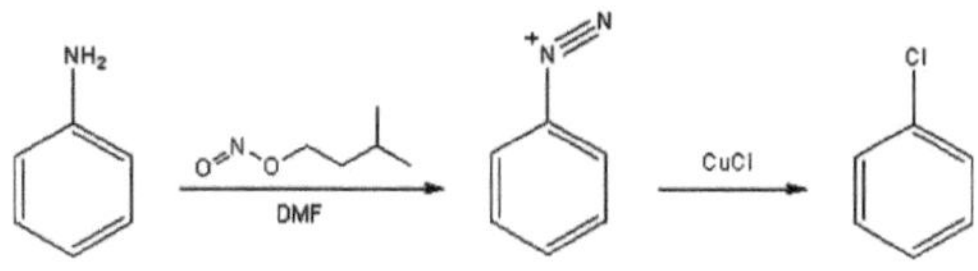

Fig. 8.10 Use of a microreactor to prepare explosive intermediates

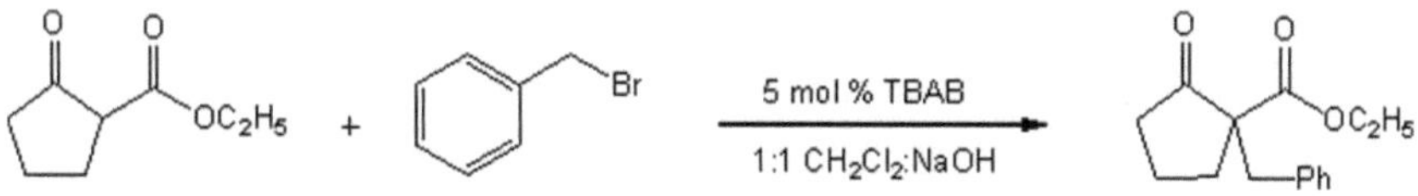

Fig. 8.11 Reaction under biphasic conditions in a microreactor

noting that the reaction was extremely efficient compared to traditional batch approaches where typical conversions of 40 % are reported.

The high SVR not only improves heat transfer, but is also advantageous when investigating biphasic processes, as illustrated by Ueno and co-workers (2003). As Fig. 8.11 illustrates, the reaction selected involved the alkylation of ethyl-2-oxocyclopentanecarboxylate with benzyl bromide in the presence of a phase transfer catalyst, tetrabutylammonium bromide (TBAB). Employing a microreactor (channel dimensions, 100 μm (wide) × 100 μm (deep) × 45 cm (length)), the authors investigated the effect of residence time on the yield of the reaction. Maintaining the reactor at room temperature, the authors investigated the effect of residence time on the conversion to product. Using a residence time of 2 min, the authors reported 75 % yield of the desired product; in comparison to a stirred batch reactor, this represented an increase of 26 % over the same reaction time. Furthermore by increasing the residence time to 10 min, an increase in yield was observed, affording the ester in 96 % yield.

Microreactor technology has also been used for reactions catalyzed by enzymes. For example using a β-galactosidase enzyme, Kanno et al. (2002) reported quantitative hydrolysis of *p*-nitrophenyl-β-D-galactopyranosideto D-galactose (Fig. 8.12) within a microreactor (channel dimensions, 200 μm (wide) × 200 μm (deep) × 40 cm (length)). The authors observed that by maintaining the microreactor at 37 °C, the reaction proceeded five times faster than in an analogous batch reaction. This enhancement was attributed to the efficient mixing obtained within the micro channel compared to a traditional batch setup.

It should also be emphasized that in addition to liquid phase reactions, microreactor technology is also highly suited to reactions involving gases. Using a silicon-glass microreactor (reaction channel, 400 μm (wide) × 400 μm (deep) × 43 cm (long)), Murphy and coworkers (2007) evaluated a series of Pd-catalyzed Heck aminocarbonylations (Fig. 8.13). Employing reactor temperatures in the range of 116–160 °C, the authors were able to efficiently convert a series of aryl halides into the respective amides in moderate to high yield (32–83 %).

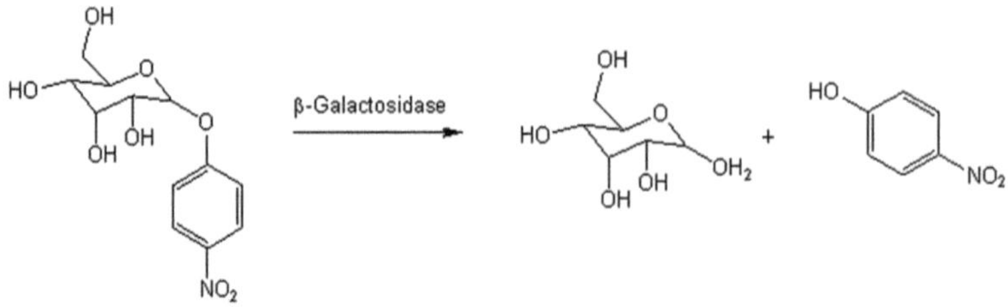

Fig. 8.12 Enzymatic catalysis in a microreactor

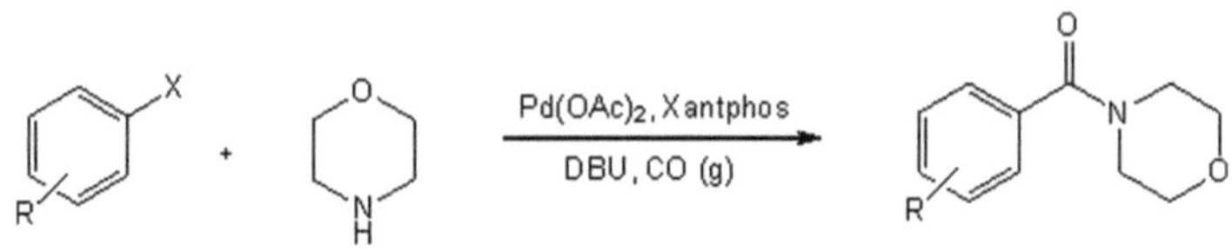

Fig. 8.13 Reaction illustrating gas phase reactions in a microreactor

8.3.2 Relevance of Microfluidics in Radiochemistry

The incorporation of radioactive nuclides into biologically relevant molecules represents a useful way to understand at a molecular level the biochemical mechanisms underlying living organisms (Serdons et al. 2009). The final image obtained through these *Nuclear Medicine* techniques allows superimposing the functional features of studied organs with anatomical details, which may be obtained by other approaches. This provides a molecular insight of the phenomena happening *in vivo*, even before they become manifested through anatomical changes (Phelps 2000). It is therefore evident that a wise utilization of this range of techniques would be reflected in high level diagnosis for a better Health Care System for general population.

It would be therefore important to have access to several of such radiolabelled molecules, generally defined as *radiopharmaceuticals*, to target different biochemical pathways (difference in lead structure) as well as for exploiting the sensitivities of different Nuclear Imaging apparatuses (difference in emission features of nuclide).

The synthesis of these radiopharmaceuticals needs to cope with three peculiar features (Cai et al. 2008), presented in Box 1. Recent reports have demonstrated that a microfluidic approach can help in coping with all these requirements (Elizarov 2009; Watts et al. 2012). Radiolabelling reactions happen with higher incorporation yields and less chemical consumption in reaction times that might be orders of magnitude smaller than the traditional vessel counterpart; in addition to that, micronized systems would allow more compact and effective shielding, together with an increased capability of automation.

Challenges of Radiochemistry

- *Small quantity of labeling agent*: the radioactivity needed for the imaging study is generally borne by a small mass of chemical species, and reactions used to incorporate such labels into desired molecular structures need to take into account this limited amount of reagent. This means that processes requiring stoichiometric (or more than stoichiometric) conditions cannot be applied to radiochemistry. This factor is generally referred as utilization of *tracer levels* and, when is not detrimental to the labeling reaction, it actually allows to reduce or even eliminate the pharmaceutical effects of the synthesized radiopharmaceutical.
- *Radiation exposure*: the use of radioactive substances implies the need of radiation shielding and reduced manual handling. This limitation is generally overcome by employing simplified chemical apparatuses (therefore not all the processes can be performed in a radiochemistry environment), by using remote handling systems, or by setting up automated synthesizers that can perform the radiosynthesis without direct human intervention. This last solution is the most desirable, but is achievable only when the processing is understood in detail and all the critical reaction parameters have been optimized thoroughly.
- *Short half-life*: some of the most useful diagnostic nuclides (e.g., ^{18}F, ^{11}C) have short half-lives, ranging from few minutes to few hours. This limitation forces the radiochemist to employ incorporation reactions that are effective in the time scale of the chosen nuclide. If this requirement is added to the reduced availability of chemical forms of the starting nuclide (e.g., ^{123}I-NaI), it is clear that the number of processes utilizable is very limited, and every mean that can speed up the rate of this small set of processes is very important.

There is therefore a big interest in integrating microfluidic systems in the synthesis of known and novel radiopharmaceuticals, especially in the case of ^{18}F and ^{11}C products, whose utilization in Positron Emission Tomography (PET) allows, more than with other imaging techniques, to investigate the molecular origins of several disease conditions. The following paragraphs summarize how micronized systems have been used in several of the production steps leading to radiopharmaceuticals.

8.3.3 Microfluidic Aided Radiolabelling Reactions

Performing the actual reaction step in the most repeatable, robust, and fast way is a very stringent and desirable feature when dealing with radiolabelling, especially with PET nuclides (e.g., ^{11}C, ^{18}F). In this field, the use of micronized systems has been proven to give access to quicker, cheaper and more reliable processes (Rensch et al. 2013). Microfluidic reactors used can be divided into 2 classes, depending on the modality of working: Micro Vessel Systems (MVS) and Micro Channel Systems (MCS) (Pascali et al. 2013).

MVS represent the direct translation of traditional vessel processes in micronized dimensions and will be discussed in Sect. 8.4. MCS, on which we will focus in the following, are based on a different approach in which the radiolabelling reaction is realized by co-flowing the reagents inside a pattern of heated micro channels (width <500 μm).

This approach has already been used for process intensification of chemical productions, and has now been applied by several groups in the production of radiopharmaceuticals. The benefits of such systems are related to the highest SVR, the ease of setup and low cost of design/testing. The high SVR allows obtaining a homogeneous heating, also on small boluses of reagents; this fact induces the most efficient heat transfer and therefore increases the kinetics of the labeling reaction.

The small channels used naturally impose to the liquids delivered a high backpressure to get through the microfluidic network, therefore allowing the use of temperatures over the atmospheric boiling points of solvents. In addition, the high SVR allows efficient cleaning of MCS by using the minimum amount of solvents. The reaction parameters are easily optimizable and upscaling is essentially performed by flowing bigger volumes of the same reagents under the same optimized conditions. The main problem of MCS is represented by the occurring of clogging in MCs, and this issue is very relevant in the case of short half-life nuclides, for which access to the microreactor and extensive cleaning might not be possible, due to radiological considerations and time constraints.

The use of a simple MCS for performing esterification reactions with small ^{11}C-alkyl and ^{18}F-fluoroalkyl reagents was the first published example of performing radiochemical reactions in microfluidics (Lu et al. 2004). Since that report, the following literature has demonstrated an increased process yield, a reduction in the chemicals used, and a general ease in setting up and accessing flexible operations. This fact has also been possible by the access to commercial systems, either designed for radiochemical reactions (http://www.scintomics.com/en/production/-mu-icr/index.html; http://www.advion.com/products/nanotek/), or dedicated to general chemistry and then adapted to radiolabelling (http://www.futurechemistry.com/home.html).

MCS have been used in the radiolabelling step in the production of several known tracers, demonstrating a higher productivity in all the cases, even starting with reduced quantities of precursor. An example of this phenomenon was reported in the radiolabelling of the dopamine D2/D3 receptors ligand ^{18}F-fallypride (see Fig. 8.14), in which the authors reported 88 % decay corrected yield (dcy) using only 1–2 % of the starting precursor needed in the vessel based procedure (Lu et al. 2009).

Another example is represented by the production of the hHSV-based reporter gene imaging agent ^{18}F-FIAU, for which the key fluorination step was realized in microfluidics, while the other 2 non-rate limiting processes were performed using

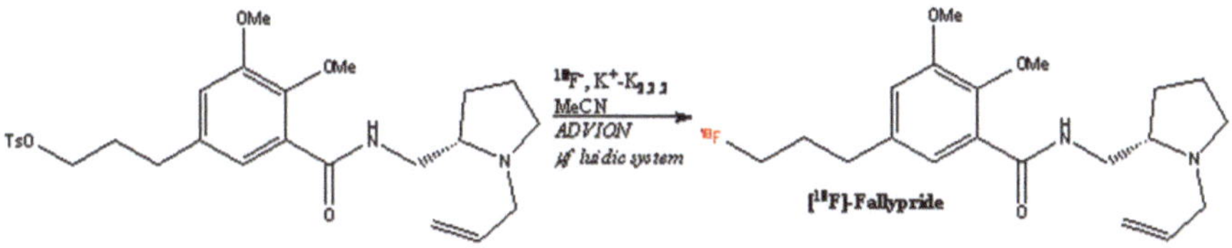

Fig. 8.14 Synthesis of [^{18}F]-Fallypride in the Advion NanoTek commercial flow system

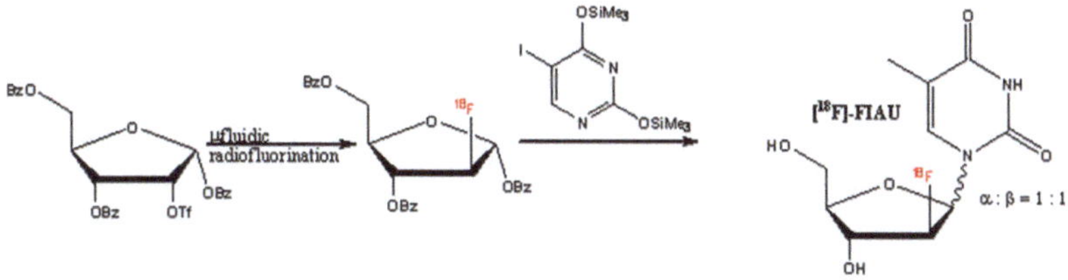

Fig. 8.15 Radiosynthesis of [^{18}F]-FIAU exploiting the high yielding microfluidic fluorination step

traditional vial reactions (Anderson et al. 2010). This allowed the authors to obtain an increased yield and purity of the final product (Fig. 8.15).

One of the useful features of PET radiopharmaceuticals is the possibility to use tracers with high Specific Activity (SA), that is the amount of radioactivity borne by the unit of mass of chemical entity (generally measured in MBq/nmol) (Lapi and Welch 2012). If SA is high enough (e.g., for receptor studies >100 MBq/nmol), the imaging study can be conducted successfully (enough radioactivity injected) without the homeostasis of the organism to be perturbed (not enough chemical mass), therefore realizing the so-called *real tracer* conditions (Fuchtner et al. 2008). SA is lowered by isotopic dilution with stable isotope, and this is generally attributed to scavenging of this element from the apparatus or the chemicals used in the process. Therefore, the possibility to use an extraordinary small quantity of precursor in microfluidics resulted useful in increasing the SA in the synthesis of ^{18}F-XeF$_2$, which can be then used as a source of electrophilic fluorine. This fluorinated synthon is produced by isotopic exchange of the stable ^{19}F-XeF$_2$ with ^{18}F-fluoride; applying microfluidic conditions allowed to increase its SA to 1.1 MBq/nmol, a level 100-fold higher than what is obtainable with other techniques (Lu and Pike 2010).

The high repeatability achievable in microfluidic MCS conditions, the possibility to use small volumes of solutions and the much faster radiolabelling, have allowed researchers to perform quick optimizations of reaction conditions, especially for the case of ^{18}F-fluorinations (Pascali et al. 2010). Traditionally, finding the best reaction conditions implies performing the most number possible of vessel reactions employing macroscopic quantities of solutions in order to be handled comfortably, reacted for 5–60 min: in the case of radiofluorinations, this means that the average experimental day will allow 5–10 different conditions to be tested and 1–10 mg of precursor to be used in every run. In MCS, the same reaction could be

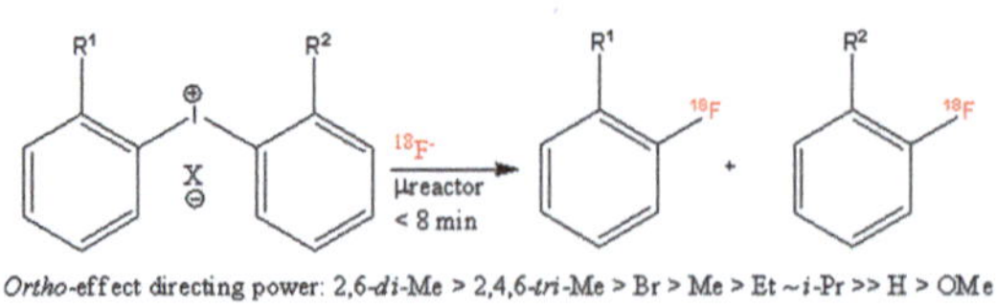

Fig. 8.16 Basic study on the effect of ortho substituents on aryliodonium salts in their radiofluorination

performed employing 20–90 s of residency time and using the same 1–10 mg of precursor but for performing 20–30 different runs. In addition to this, it must be pointed out that the optimized microfluidic processes have a higher replicability intra- and inter-laboratory, due to the simple setup of MCS. Finally, the passage from test reactions, using small quantity of activity, to real production runs, using the entire radioactivity needed, is easily obtainable by flowing the desired quantities of solutions in the same optimized system.

Several groups have exploited this feature for studying basic radiochemistry reactions, as in the case of the use of different aryliodonium salts as precursors for difficult ^{18}F-fluoroaromatics. In this case, the use of fast microfluidic conditions allowed unveiling an unprecedented *ortho* effect (Fig. 8.16) in the labeling of such compounds (Chun et al. 2010).

In other fields, the utility of MCS has been demonstrated when very expensive precursors need to be used, as in the case of macromolecule labeling. In one case, preformed ^{18}F-succinimidylbenzoate (^{18}F-SFB) has been used to label on the lysine residue two different polypeptides (Richter et al. 2012), demonstrating that the microfluidic approach allows a quicker optimization using a much smaller amount of synthetic precursor; in addition, the incorporation yields experienced were higher when compared to similar processes conducted in an Eppendorf vial. In another case, a droplet system implemented on a custom-built MCS allowed the researchers to perform up to 125 different labeling experiments using in total 5 μL of protein solution (5 mg/mL) (Liu et al. 2011).

Another application of the high repeatability and of the small volumes utilizable in MCS relays in the possibility to perform Dose On Demand (DOD) productions. By this concept, it is intended the use of the minimum quantity of radioactive agent, precursor and solvents to obtain, in the shortest time-frame possible, just one dose of radiopharmaceutical ready for injection. This would be a game-changer in the current PET tracers' production scenario, for which the whole lot of starting activity and precursor is used to obtain a whole big batch of final product that is then divided in different doses. The DOD approach would, in theory, give an increased flexibility to the final user, also in terms of obtaining different tracers with the same starting activity. The proof of principle of this system has been demonstrated in the synthesis of several ^{18}F-fluoroalkylcholine analogues (Fig. 8.17), which might be used for tumor imaging (Pascali et al. 2011).

This represented also the first example of a successful 2-step/1-flow production process, currently reported only when the second step was a not rate-limiting one (e.g., hydrolysis).

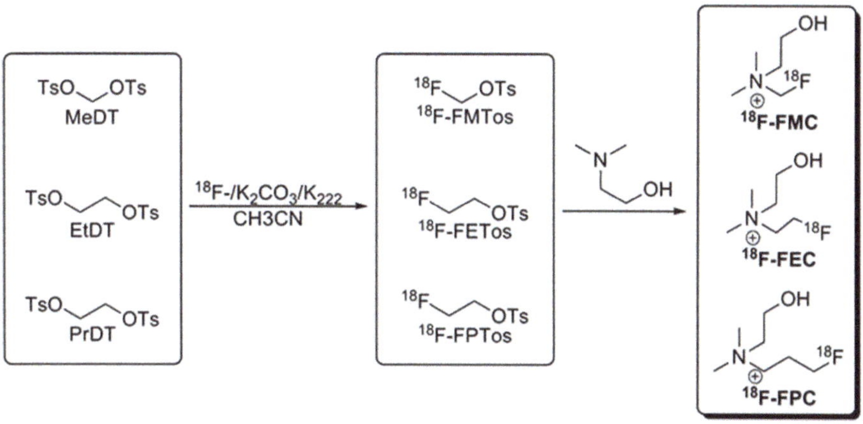

Fig. 8.17 [^{18}F]-Fluoroalkylcholines produced in commercial microfluidic system following a DOD approach

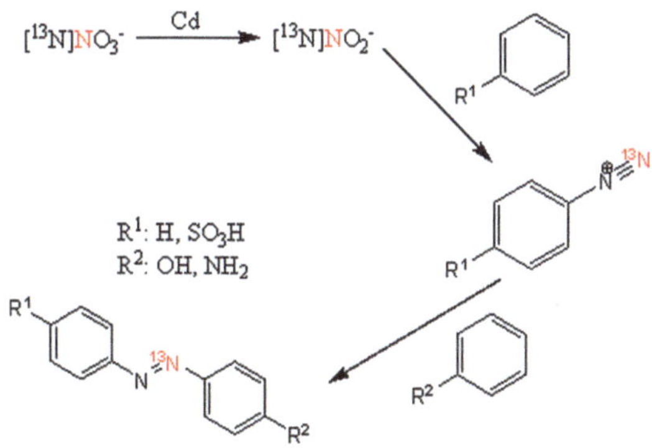

Fig. 8.18 Study for 2-step production of azo-compounds using microfluidic system with short-lived ^{13}N

Since the use of microfluidics has demonstrated reduced reaction times, a natural outcome would be its use with very short half-life nuclides. In fact, MCS have been successfully employed in custom adapted gas-liquid apparatuses for performing Pd-catalyzed ^{11}C-carbonylation reactions (Miller et al. 2011). Other homogeneous ^{11}C-carbonylations have also been implemented in MCS, using a Cu complex as ^{11}C-CO source; in this case the process yield was not statistically higher, but the process resulted to be safer in terms of confinement of radioactive exhaust gases (Kealey et al. 2011).

^{13}N is a very short half-life PET nuclide ($t_{1/2} = 10$ min) and its use is very limited due to this feature, even if it would allow a mimetic labeling of well characterized biomolecules. MCS have been used for synthesizing in flow several ^{13}N-nitrosothiols and ^{13}N-azacompounds (Fig. 8.18), and they were obtained utilizing very short residency times (<2 min) and in higher yields than the corresponding traditional vessel methods (Gaja et al. 2012).

8.3.4 Microfluidics in Radiochemistry Integrated Systems

After the proper labelling reaction, other secondary reactions can be performed (deprotection, hydrolysis), and they can be done in a microfluidic fashion as well. An interesting approach was recently developed by interfacing two commercial flow systems, one performing the microfluidic radiolabelling and the second achieving a flow hydrogenation on a catalytic bed (Liang et al. 2013). In this way, it was possible to obtain a very quick and effective benzyl group deprotection, thus providing a direct route to further investigations on CABS513 (Fig. 8.19), a prospective Alzheimer PET imaging agent.

However, even if the radiolabelling step is a central one in the process of production of PET tracers, additional steps like purification and formulation or, in the case of ^{18}F radiochemistry, concentration and activation, would also benefit from the speed, reliability and robustness attainable using a microfluidic approach.

In most of the PET tracers' production, the raw starting nuclide generally needs to be transformed into a chemical form that allows the radiolabelling reaction to occur. For example, ^{11}C is generally produced at a biomedical cyclotron as small bolus of $^{11}C\text{-}CO_2$ in a gaseous stream; this reagent generally needs to be concentrated in a small volume and transformed in more reactive forms, e.g., $^{11}C\text{-}CH_3I$, ^{11}C-HCN, $^{11}C\text{-}CH_3TfO$ (Miller et al. 2008). Unfortunately, no microfluidic approaches have been reported up to date in order to perform these reactions in a more efficient way.

On the other hand, the activation of ^{18}F-fluoride, which is generated at a cyclotron as aqueous solution, has been subjected to some microfluidic improvements. The presence of excess water prohibits the use of this anion in nucleophilic substitution reactions; activation is therefore required, in order to render such species utilizable in the synthesis of ^{18}F-fluorinated PET radiopharmaceuticals.

Traditionally, the ^{18}F-fluoride produced is trapped into an anion exchange resin; the excess water is rinsed out with organic polar aprotic solvent and the radioactive nuclide is subsequently released using a solution of base (e.g., K_2CO_3) and kryptate ligand (typically $K_{2.2.2}$) in an organic solution containing the minimum amount of water. The eluted solution might be further dried by means of azeotropic distillation for reducing the water content to an even lower value (Cai et al. 2008).

A useful approach for activation of ^{18}F-fluoride is represented by electrochemical systems (Saito et al. 2007; Hamacher et al. 2002). In these applications, the trapping of the anion is realized by applying a suitable electric potential, which is maintained also while rinsing the system with a drying solvent. The releasing is obtained by reversing the potential and passing a solution containing the previously

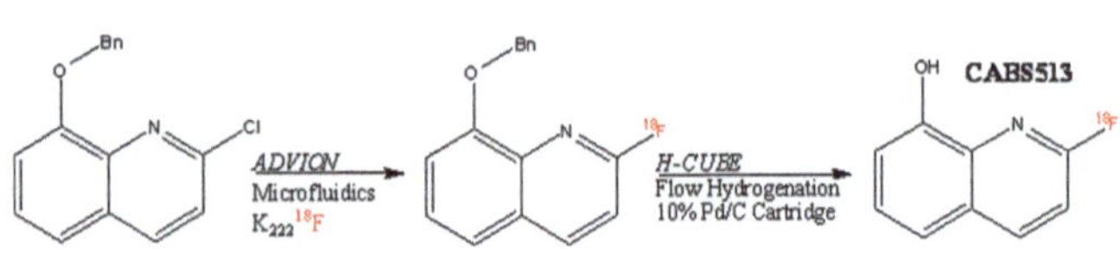

Fig. 8.19 Production of CABS513 using two subsequent flow reactions

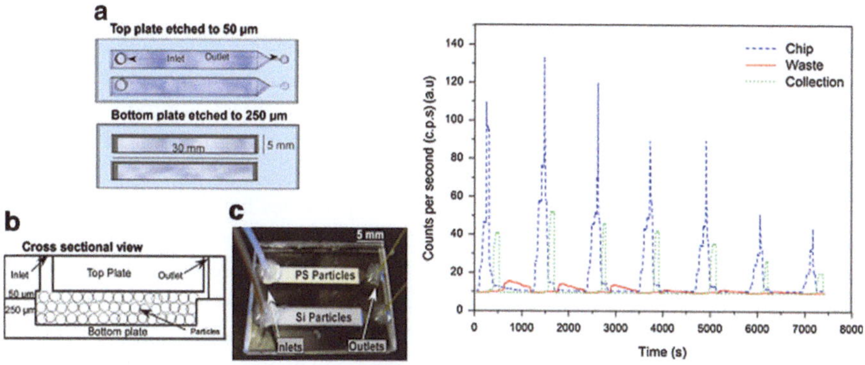

Fig. 8.20 Micro-SPE system and its use in trapping radioactive fluoride into anion exchange beads

cited activating agents. The implementation of this concept into a microfluidic device was reasonably straightforward and such systems have been successfully used for producing radiofluorination complexes active in the following reaction. However, anode issues have been underlined, as the trapping and especially the release phase strongly depends on the nature of this electrode material. Glassy carbon anodes, used in the first prototypes, suffer from surface degradation and low repeatability of trap & release cycle; a set of metal anodes has been recently tested, and brass has been selected as the material of choice (Sadeghi et al. 2013). However, also in this case, a noticeable reduction of the releasing efficacy over the usage has been reported.

The use of micro-SPE systems has also been tested to verify whether the reduction of dimensions and resin mass would lead to increased reliability and potential reutilizability (Fig. 8.20).

The mass of resin contained in the traditional anion-exchange cartridges is > 100 mg and has a wide particle size distribution (e.g., Waters, QMA Light); however, due to the tracer levels of the ^{18}F-fluoride, a reduced quantity and a better choice of the used particles should allow the same efficiency in trapping the radioactive anion and releasing it in a very small volume. Therefore, some producers are now supplying short 1/16″ PTFE tubing filled with <10 mg of suitable resin (MP1 cartridge, ORTG, TN, USA), and this approach has proved reliable in several applications. Further to this, a dedicated design of a micronized chamber has been reported, and allowed to realize not only a suitable efficiency but also an effective recycling of the resin (De Leonardis et al. 2011).

This particular micronized module allowed repeating the trap & release cycle up to 40 times without loss in performance, thus envisaging the possibility to restart a full production cycle of an ^{18}F-fluorinated tracer without direct exposure of the operator for exchanging the resin.

Specifically designed micro-SPE has been used also for the purification step of the blockbuster PET tracer ^{18}F-FDG (Tarn et al. 2013). In this case, a simple filtration through a set of resins is needed to obtain a solution free of contaminants.

Available literature has demonstrated that this approach might be feasible for producing a stream of purified FDG out of a continuously running MCS apparatus. This prototype actually reported the integration of the whole production process in a very compact flow system that generated a continuous stream of product, starting from a continuous feed of radiofluorination complex and chemicals (Arima et al. 2013) (Fig. 8.21).

Unfortunately, no other microfluidic based approaches have been employed for purifying radiopharmaceuticals. The most general and utilized route relies on dilution with an appropriate solvent (e.g., water, buffer) of the microfluidically produced reaction mixture, single-pass semi-preparative HPLC, peak collection and SPE workup (e.g., C-18) for obtaining a suitable formulated injectable in an EtOH/saline mixture. In some cases such processes can be obtained by interfacing with traditional radiosynthetic boxes; in other cases, it was possible to integrate and control additional hardware directly from the microfluidic system (Pascali et al. 2014). In both cases, it is important the processes following the radiolabelling reaction to be custom modified for the modalities and peculiarities of microfluidic systems.

8.4 Beyond Planar Microchannel-Based Microreactors

After describing planar MC-based microreactors and their applications to organic chemistry and radiochemistry, we will give a short overview on *meso* flow reactors (where the characteristic dimensions are slightly larger) which represent all the way to production and micro vessel systems (MVS) for radiopharmaceutical synthesis.

8.4.1 Mesoreactors and Production

As outlined, one advantage of microreactors is that heat transfer is several orders of magnitude more efficient than that of batch reactors. Although this is clearly advantageous for many reactions this level of process intensification is not always necessary for every reaction and as a result a significant number of companies are now marketing laboratory apparatus containing tubular reactors which have millimeter-sized reaction channels with volumes typically ranging from 1 to 20 mL (Fig. 8.22).

Azides represent a synthetically useful functional group however the hazards associated with their preparation have limited their production and use. Kopach reported (Kopach et al. 2009) the development of a continuous flow reactor suitable for the synthesis of azide from the chloride (Fig. 8.23). Using a stainless steel tube reactor (dimensions, 1.59 mm (o.d.) × 0.64 mm (i.d.) × 63.1 m (long), volume, 20 mL), the authors investigated the reaction at a series of temperatures. The authors identified 90 °C as the optimal temperature at a residence time of 20 min; affording azide in 97 % conversion. Operating the reactor continuously for 2.8 h

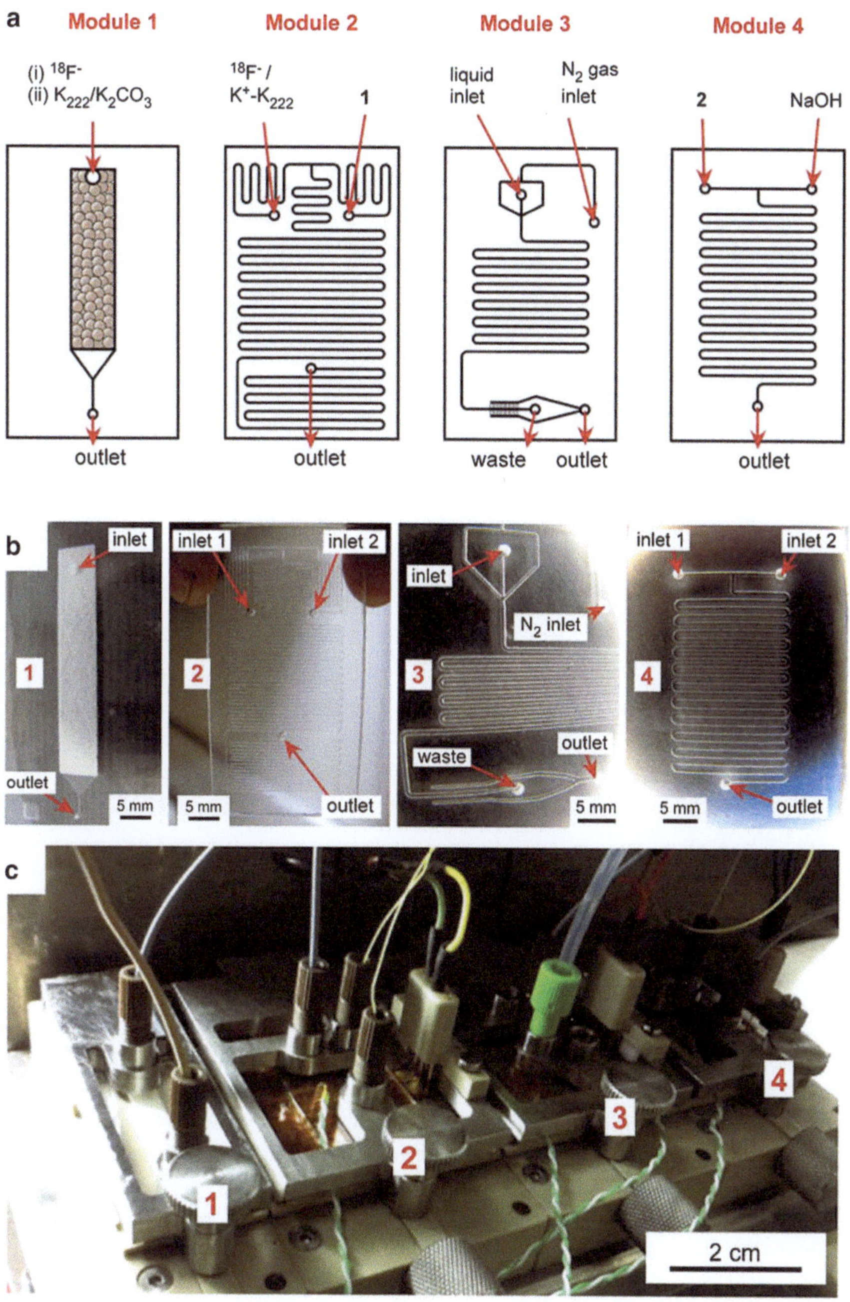

Fig. 8.21 Automated prototype for fully microfluidic continuous production [^{18}F]-FDG

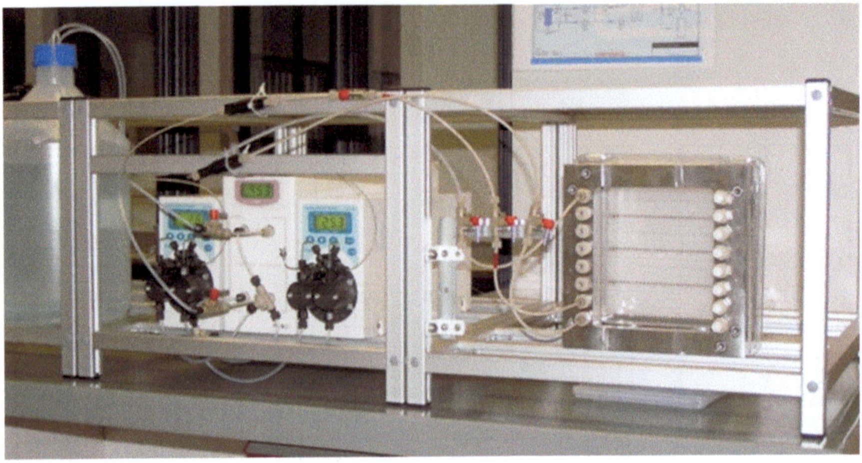

Fig. 8.22 A mesoreactor system used for the production under continuous flow

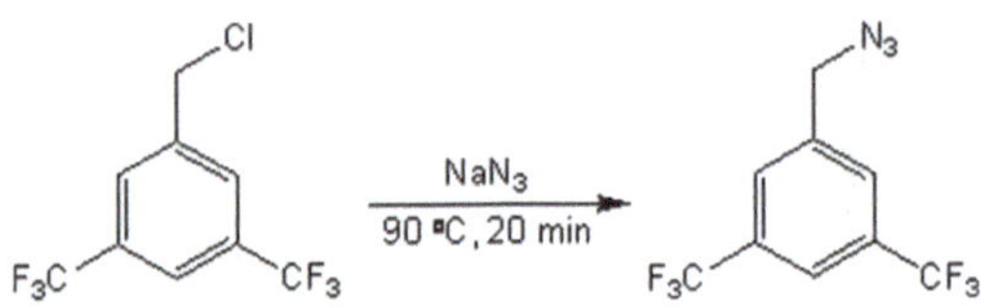

Fig. 8.23 Azide synthesis in a meso scale reactor

enabled the authors to produce 25 g of the azide in 94 % isolated yield after extraction.

Introduction of fluorine into pharmaceuticals is very important and as such efficient synthetic methods are sought for the introduction of fluorine into small organic compounds. A recent example reported by Gustafson et al. (2008) demonstrated the efficient synthesis of fluorinated alcohols, aldehydes and carboxylic acids using diethylaminosulphurtrifluoride (DAST) in a PTFE tube reactor (Volume, 16 mL). Employing a reaction time of 16 min and a temperature of 70 °C, the authors were able to isolate the target fluorinated compound in yields ranging from 40 to 100 % (Fig. 8.24).

Employing a commercially available tubular reactor (Vapourtec), Riva et al. (Riva et al. 2010) demonstrated the reaction of a series of carbonyl containing compounds with Grignard reagents as a means of synthesizing substituted alcohols. Using the developed methodology, the authors extended their investigation to evaluate the synthesis of (rac)-Tramadol (Fig. 8.25). Using the reactor, the authors mixed ketone (0.25 M) with Grignard reagent (1.2 eq.) at room temperature for a period of 33 min. Under these conditions, the authors were able to isolate the target compound in 96 % yield as a diastereomeric mixture (8:2). The high degree of reaction control obtained under flow conditions has been shown to be advantageous

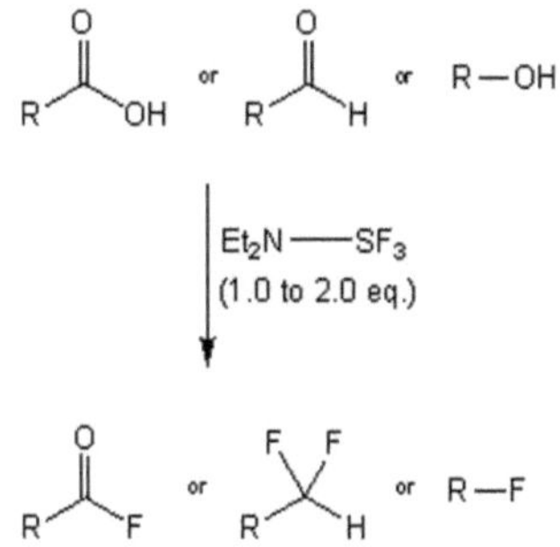

Fig. 8.24 Fluorinations in meso scale reactors

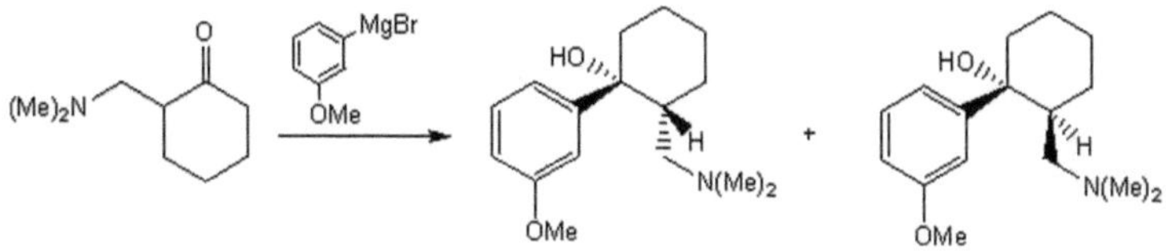

Fig. 8.25 Grignard reaction in a mesoreactor

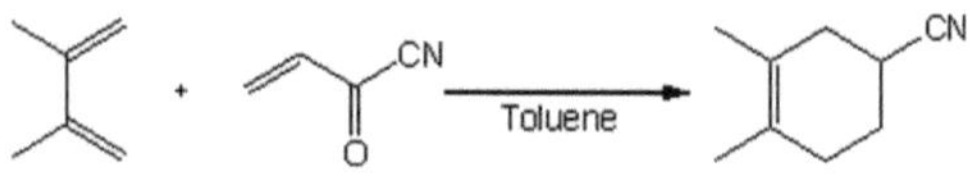

Fig. 8.26 High temperature synthesis in a mesoreactor

in particular for the Grignard reaction, as a means of suppressing side reactions and increasing product purity.

Utilizing a high temperature (300 °C) and pressure (200 bar) stainless steel tubular reactor, Razzaq et al. (2009) were able to expand the processing window usually employed by research chemists. Employing the cycloid addition of 2,3-dimethylbutadiene and acrylonitrile as a suitable reaction (Fig. 8.26), the authors were able to demonstrate dramatic reductions in reaction time as a result of performing the reaction under continuous flow. Using toluene as the solvent, the authors observed complete conversion to 3,4-dimethylcyclohex-3-enecarbonitrile with quantitative conversion obtained at 250 °C (200 bar) and a reaction time of 5 min.

Transition metal catalyzed C–C bond forming reactions form one of the most widely studied classes of reaction that have been performed under continuous flow conditions, with a recent review by Singh et al. (Singh et al. 2008) providing background into the types of reactions investigated. As an example, Wilson et al. (2004) demonstrated a Suzuki coupling reaction in a borosilicate glass coil reactor (Volume, 10 mL). Using EtOH as the reaction solvent and triethylamine as the base, the authors investigated the synthesis of 4-(benzofuran-2-yl)benzaldehyde *via* the $PdCl_2(PPh_3)_2$ catalyzed coupling of 4-bromobenzaldehyde with benzofuran-

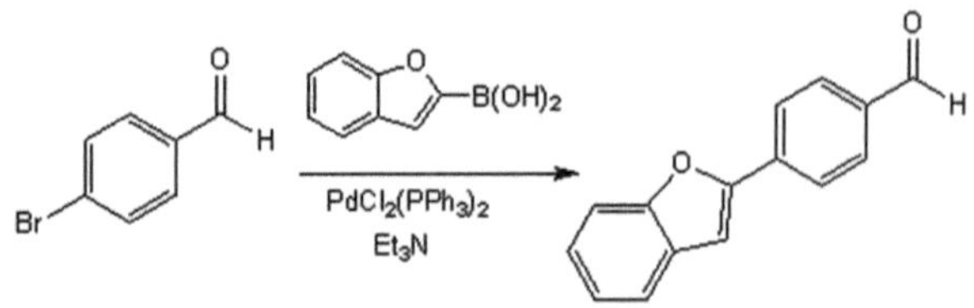

Fig. 8.27 Catalysis in a mesoreactor

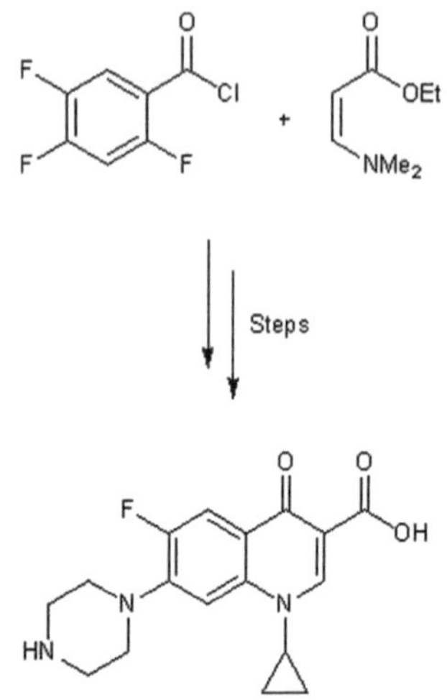

Fig. 8.28 Total synthesis of API in flow reactor

2-yl boronic acid (Fig. 8.27). Optimal conditions were found to be a flow rate of 0.25 mL/min (residence time of 8 min) and a reactor temperature of 140 °C, affording the coupling product in 84 % yield.

The previous examples have been selected to illustrate how important synthetic transformations have been transferred to flow reactors. However it needs to be emphasized that more recently there has been an increased trend to directly couple micro and flow reactors together in order to enable the efficient multistep synthesis of complex products. An early example of a multistep reaction performed under continuous flow conditions was the Ciprofloxacin synthesis reported by Schwalbe (Fig. 8.28) (Schwalbe et al. 2002).

During their research into the development of new reaction pathways, McQuade and coworkers (Bogdan et al. 2009) developed a continuous synthetic route for the preparation of the analgesic Ibuprofen (Fig. 8.29). Using a PFA tube reactor, the authors devised a pathway comprising of a Friedel–Crafts acylation, followed by a 1,2-aryl migration and finally ester hydrolysis to furnish the target compound in 68 % yield and 96 % purity. Purification was subsequently performed resulting in a 51 % overall yield.

Employing a series of polymer-supported reagents and catalysts, Ley and coworkers (Hopkin et al. 2010) have demonstrated the continuous flow synthesis of a series of pharmaceutically relevant compounds, with the synthesis of Imatinib being one example (Fig. 8.30). Performing three discrete reaction steps, the authors

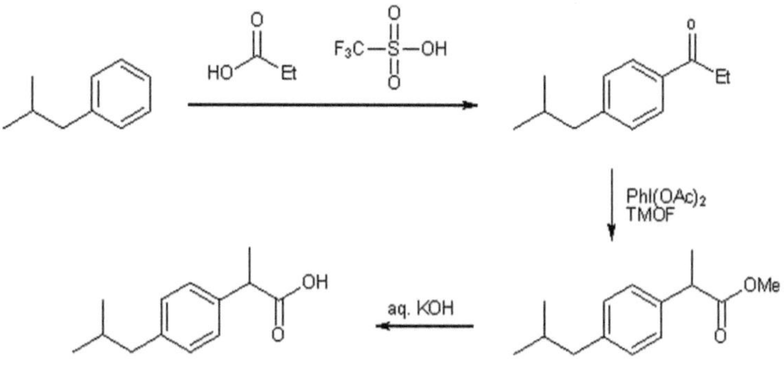

Fig. 8.29 Synthesis of ibuprofen in a mesoreactor

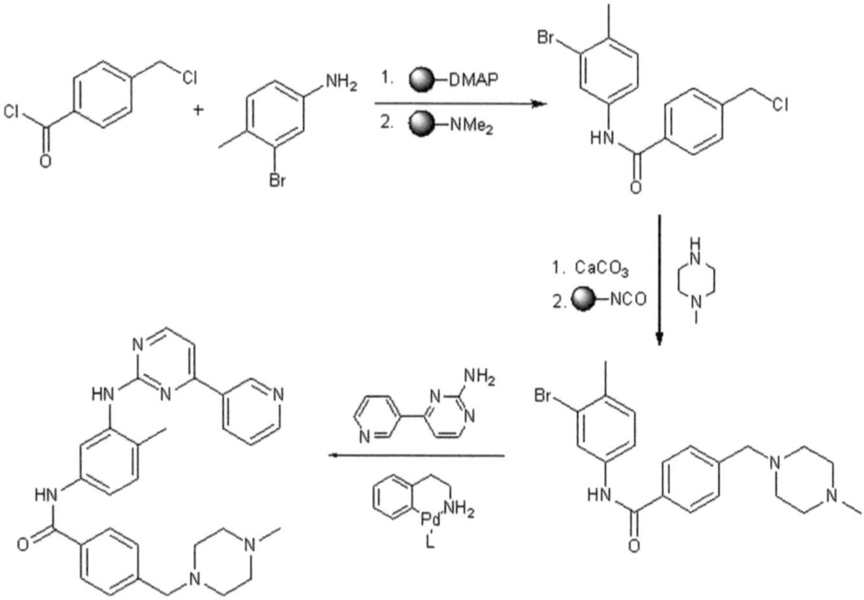

Fig. 8.30 Use of solid supported reagents in total synthesis in flow reactors

were able to isolate the final active pharmaceutical ingredient (API) in an overall yield of 32 %.

As outlined, a major problem observed with conventional batch technology is the failure to scale-up successful reactions; this is particularly problematic for highly exothermic processes. In industry this often means redevelopment of synthetic routes on several occasions when transferring from laboratory, to pilot scale and then finally to production. This is a both a time consuming and costly process. This is where flow processing has major advantages as the reactor length and volume can be flexibly adjusted to enable production at the relevant scale.

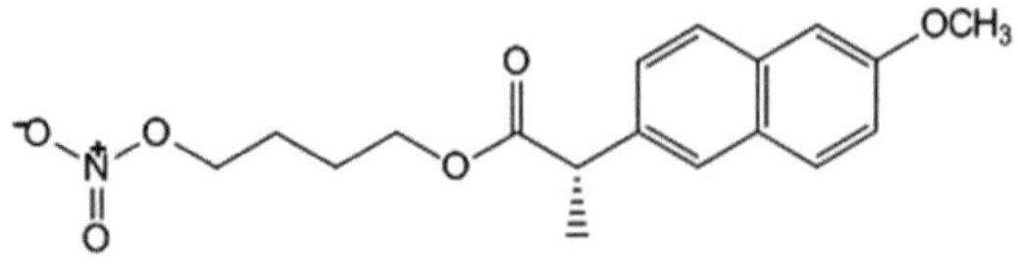

Fig. 8.31 Naproxen derivative

In an example illustrating incredible processing safety associated with flow reactor methodology, Reintjens and coworkers (Thayer 2005) demonstrated selective nitration using neat nitric acid. Synthesizing a Naproxen derivative (Fig. 8.31), the researchers report that a 150 mL reactor was used to generate the target compound at a rate of 13 kg/h, with intrinsic safely levels not attainable using conventional batch methodology. Operating eight reactors in parallel enabled the scale to be increased to 100 kg/h which equates to a theoretical throughput of 800 tonnes/annum. The investigation illustrates the ease with which flow processes can be scaled to achieve production-scale quantities whilst retaining a relatively small system footprint.

In summary, it is now well established that microfluidic and flow reactor technology enables reactions to be performed more rapidly, efficiently, and selectively than batch reactions using very small amounts of reagent. By optimizing the residence time and temperature within the reactor, chemists are able to perform reactions that are very difficult to control in batch.

8.4.2 Micro Vessel Systems Approach to Radiochemistry

Micro Vessel Systems (MVS) represent the direct translation of traditional vessel processes in micronized dimensions; therefore, while macro-vessel reactors are characterized by volumes in the 2–5 mL range, MVS features microreactors in the 10 nL–50 μL range. These volumes allow obtaining a higher localized reagent concentration, faster batch processes (e.g., evaporation), application of optimization paradigms derived from macro-vessel approach. However, these systems have higher cost of prototyping and testing; dedicated delivery systems of minute quantities of liquids, microheaters, materials, geometry, and interfacing are needed in each case and can be difficultly translated from one prototype to the other (Elizarov et al. 2010). Even if the use of MVS revealed to be a groundbreaking application, which allowed reporting the synthesis of ^{18}F-FDG on the smallest radiochemical integrated device possible (Lee et al. 2005), their use has been limited up to now. Some interesting works reported the use of MVS made of plastic materials (polydimethylsiloxane, polycarbonate) to handle not only ^{18}F (Gillies et al. 2006a, b) and ^{68}Ga (short half-life) (Zeng et al. 2013), but also ^{124}I, ^{99m}Tc (Gillies et al. 2006a, b) and ^{64}Cu (longer half-life) (Wheeler et al. 2010); the latter systems were tested in the view of using cheap single use reactors that may be discarded into radioactive waste collection, and therefore minimizing the risk of contamination from longer lived isotopes. An important milestone that was reached with MVS approach was the production of ^{18}F-Fallypride for using into PET human studies (Lebedev et al. 2013). This fact has a fundamental impact on the clinical world, since producing radiopharmaceuticals for human studies requires a high level of control, generally referred to Good Manufacturing Practice (GMP). This report demonstrated the ability for a microfluidic system to respect these stringent requirements using a MVS approach, which practically mimics traditional approaches to a smaller scale.

A MVS built with a new approach, was the one using the ElectroWetting On Dielectric (EWOD) principle (Keng et al. 2012). This system allowed moving liquids exploiting their polarity and the difference in electric potential across neighboring microareas. In this way, it was possible to dispense very minute amounts of reagents (tenths of nL) in a very precise manner without the use of any mechanically moving parts. Heating was also electrically operated on the reactor zone, and batch operations were quick and efficient. This prototype has been used with success for the synthesis of several ^{18}F-fluorinated tracers, even using 2-step/1-pot processes (Chen et al. 2012).

MVS systems reported up to date have not been used for performing reactions at high pressure (>5 bars) or high temperature (>150 °C), mainly due to the difficulty in realizing at this micronized scale valves and containers that can withstand these conditions. A very recent approach has solved this problem by the use of phase-change valves. In this work, a PTFE loop has been loaded with the reactions mix (31 μL) and placed onto a heating element; each of the two extremities of the tubing have been filled with a small inert gas gaps (0.2 μL) and a volume of DMSO (3.2 μL). These extremities have been placed at −19 °C, thus realizing the freezing of DMSO, while the core of the loop containing the reagents in CH_3CN has been heated at 165 °C for a set time. This system allowed to obtain 83 % incorporation yield for ^{18}F-FAC (Fig. 8.32), a tracer for which high pressure and high temperature conditions are needed to obtain useful yields (Ma et al. 2014).

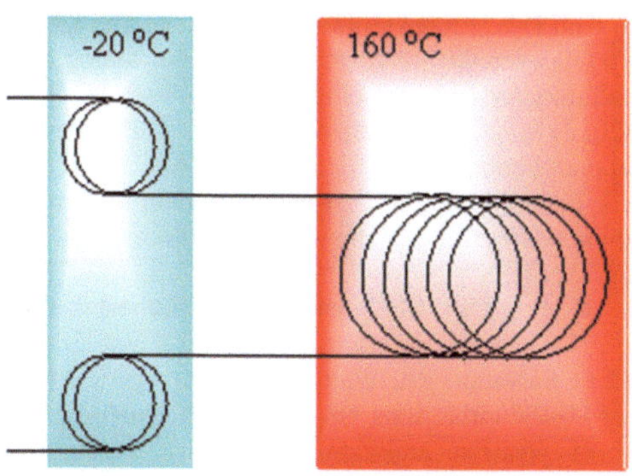

Fig. 8.32 Implementation of phase-change valves, employing frozen DMSO plugs to perform high temperature and high pressure radiofluorinations

8.5 Conclusion

This chapter is intended to give an overview on microchannel-structured planar microreactors and their application to organic chemistry. From a builder point of view, the geometries and the design constraints are reported for the microreactors most commonly used in the research field. The application of these devices to the synthesis of several chemical compounds is reported, with a particular care towards radiosynthesis that is considered as a field in which microfluidics can be easily translated from pure research to the production level.

References

Anderson H, Pillarsetty N, Cantorias M, Lewis JS (2010) Improved synthesis of 2′-deoxy-2′-[18F]-fluoro-1-beta-D-arabinofuranosyl-5-iodouracil ([18F]-FIAU). Nucl Med Biol 37 (4):439–442

Arima V, Pascali G, Lade O, Kretschmer HR, Bernsdorf I, Hammond V, Watts P, De Leonardis F, Tarn MD, Pamme N (2013) Radiochemistry on chip: towards dose-on-demand synthesis of PET radiopharmaceuticals. Lab Chip 13(12):2328–2336

Bej SK, Dabral RP, Gupta PC, Mittal KK, Sen GS, Kapoor VK, Dalai AK (2000) Studies on the performance of a microscale trickle bed reactor using different sizes of diluent. Energ Fuel 14 (3):701–705

Bej SK, Dalai AK, Maity SK (2001) Effect of diluent size on the performance of a micro-scale fixed bed multiphase reactor in up flow and down flow modes of operation. Catal Today 64 (3–4):333–345

Bogdan AR, Poe SL, Kubis DC, Broadwater SJ, McQuade DT (2009) The continuous-flow synthesis of ibuprofen. Angew Chem Int Ed 48(45):8547–8550

Cai L, Lu S, Pike V (2008) Chemistry with [18F]fluoride ion. Eur J Org Chem 17:2853–2873

Causon TJ, Broeckhoven K, Hilder EF, Shellie RA, Desmet G, Eeltink S (2011) Kinetic performance optimisation for liquid chromatography: principles and practice. J Sep Sci 34 (8):877–887

Cortes-Quiroz CA, Azarbadegan A, Zangeneh M, Goto A (2010) Analysis and multi-criteria design optimization of geometric characteristics of grooved micromixer. Chem Eng J 160 (3):852–864

Chen S, Javed R, Lei J, Kim H-K, Flores G, Dam RMV, Keng PY, Kim C-J (2012) Synthesis of diverse tracers on EWOD microdevice for positron emission tomography (PET). In: Solid-state sensors, actuators and microsystems workshop, Hilton Island, SC, 3–7 June 2012

Chun J-H, Pike VW (2012) Single-step radiosynthesis of "18F-Labeled Click Synthons" from azide-functionalized diaryliodonium salts. Eur J Org Chem 2012(24):4541–4547

Chun JH, Lu S, Lee YS, Pike VW (2010) Fast and high-yield microreactor syntheses of ortho-substituted [(18)F]fluoroarenes from reactions of [(18)F]fluoride ion with diaryliodonium salts. J Org Chem 75(10):3332–3338

Chambers RD, Spink RCH (1999) Microreactors for elemental fluorine. Chem Commun 10:883–884

De Leonardis F, Pascali G, Salvadori PA, Watts P, Pamme N (2011) On-chip pre-concentration and complexation of [18F]fluoride ions via regenerable anion exchange particles for radiochemical synthesis of positron emission tomography tracers. J Chromatogr A 1218 (29):4714–4719

Elizarov AM (2009) Microreactors for radiopharmaceutical synthesis. Lab Chip 9(10):1326–1333

Elizarov AM, van Dam RM, Shin YS, Kolb HC, Padgett HC, Stout D, Shu J, Huang J, Daridon A, Heath JR (2010) Design and optimization of coin-shaped microreactor chips for PET radiopharmaceutical synthesis. J Nucl Med 51(2):282–287

Erickson D, Li D (2002) Influence of surface heterogeneity on electrokinetically driven microfluidic mixing. Langmuir 18(5):1883–1892

Fang H, Xiao Q, Wu F, Floreancig PE, Weber SG (2010) Rapid catalyst screening by a continuous-flow microreactor interfaced with ultra-high-pressure liquid chromatography. J Org Chem 75 (16):5619–5626

Fortt R, Wootton RCR, de Mello AJ (2003) Continuous-flow generation of anhydrous diazonium species: monolithic microfluidic reactors for the chemistry of unstable intermediates. Org Process Res Dev 7(5):762–768

Fuchtner F, Preusche S, Mading P, Zessin J, Steinbach J (2008) Factors affecting the specific activity of [18F]fluoride from a [18O]water target. Nuklearmedizin 47(3):116–119

Gaja V, Gómez-Vallejo V, Cuadrado-Tejedor M, Borrell JI, Llop J (2012) Synthesis of 13N-labelled radiotracers by using microfluidic technology. J Labelled Comp Rad 55 (9):332–338

Gillies JM, Prenant C, Chimon GN, Smethurst GJ, Dekker BA, Zweit J (2006a) Microfluidic technology for PET radiochemistry. Appl Radiat Isot 64(3):333–336

Gillies JM, Prenant C, Chimon GN, Smethurst GJ, Perrie W, Hamblett I, Dekker B, Zweit J (2006b) Microfluidic reactor for the radiosynthesis of PET radiotracers. Appl Radiat Isot 64 (3):325–332

Gobby D, Angeli P, Gavriilidis A (2001) Mixing characteristics of T-type microfluidic mixers. J Micromech Microeng 11(2):126

Gustafsson T, Gilmour R, Seeberger PH (2008) Fluorination reactions in microreactors. Chem Commun 26:3022–3024

Hamacher K, Hirschfelder T, Coenen HH (2002) Electrochemical cell for separation of [18F] fluoride from irradiated 18O-water and subsequent no carrier added nucleophilic fluorination. Appl Radiat Isot 56(3):519–523

Hardt S, Schönfeld F (2003) Laminar mixing in different interdigital micromixers: II. Numerical simulations. AIChE J 49(3):578–584

Hartman RL (2012) Managing solids in microreactors for the upstream continuous processing of fine chemicals. Org Process Res Dev 16(5):870–887

Haselhuhn F, Kind M (2003) Pseudo-polymorphic behavior of precipitated calcium oxalate. Chem Eng Technol 26(3):347–353

Hessel V, Hardt S, Löwe H, Schönfeld F (2003) Laminar mixing in different interdigital micromixers: I. Experimental characterization. AIChE J 49(3):566–577

Hessel V, Hofmann C, Löb P, Löhndorf J, Löwe H, Ziogas A (2005a) Aqueous Kolbe–Schmitt synthesis using resorcinol in a microreactor laboratory rig under high-p, T conditions. Org Process Res Dev 9(4):479–489

Hessel V, Löwe H, Schönfeld F (2005b) Micromixers – a review on passive and active mixing principles. Chem Eng Sci 60(8–9):2479–2501

Hopkin MD, Baxendale IR, Ley SV (2010) A flow-based synthesis of Imatinib: the API of Gleevec. Chem Commun 46(14):2450–2452

Hossain S, Ansari MA, Kim K-Y (2009) Evaluation of the mixing performance of three passive micromixers. Chem Eng J 150(2–3):492–501

http://www.advion.com/products/nanotek/

http://www.futurechemistry.com/home.html

http://www.scintomics.com/en/production/-mu-icr/index.html

Iwasaki T, Yoshida JI (2005) Free radical polymerization in microreactors. Significant improvement in molecular weight distribution control. Macromolecules 38(4):1159–1163

Jung YJ, Park SH, Song KH, Choe J (2012) Recrystallization of polyethylene submicron particles using a continuous flow micromixer system. Powder Technol 217:325–329

Kanno K-I, Maeda H, Izumo S, Ikuno M, Takeshita K, Tashiro A, Fujii M (2002) Rapid enzymatic transglycosylation and oligosaccharide synthesis in a microchip reactor. Lab Chip 2(1):15–18

Kealey S, Plisson C, Collier TL, Long NJ, Husbands SM, Martarello L, Gee AD (2011) Microfluidic reactions using [11C]carbon monoxide solutions for the synthesis of a positron emission tomography radiotracer. Org Biomol Chem 9(9):3313–3319

Keng PY, Chen S, Ding H, Sadeghi S, Shah GJ, Dooraghi A, Phelps ME, Satyamurthy N, Chatziioannou AF, Kim CJ, van Dam RM (2012) Micro-chemical synthesis of molecular probes on an electronic microfluidic device. Proc Natl Acad Sci U S A 109(3):690–695

Kim DJ, Oh HJ, Park TH, Choo JB, Lee SH (2005) An easily integrative and efficient micromixer and its application to the spectroscopic detection of glucose-catalyst reactions. Analyst 130 (3):293–298

Kim DS, Lee SW, Kwon TH, Lee SS (2004) A barrier embedded chaotic micromixer. J Micromech Microeng 14(6):798

Kopach ME, Murray MM, Braden TM, Kobierski ME, Williams OL (2009) Improved synthesis of 1-(azidomethyl)-3,5-bis-(trifluoromethyl)benzene: development of batch and microflow azide processes. Org Process Res Dev 13(2):152–160

Kutter JP (2012) Liquid phase chromatography on microchips. J Chromatogr A 1221:72–82

Lapi SE, Welch MJ (2012) A historical perspective on the specific activity of radiopharmaceuticals: what have we learned in the 35 years of the ISRC? Nucl Med Biol 39(5):601–608

Lebedev A, Miraghaie R, Kotta K, Ball CE, Zhang J, Buchsbaum MS, Kolb HC, Elizarov A (2013) Batch-reactor microfluidic device: first human use of a microfluidically produced PET radiotracer. Lab Chip 13(1):136–145

Lee C-C, Sui G, Elizarov A, Shu CJ, Shin Y-S, Dooley AN, Huang J, Daridon A, Wyatt P, Stout D, Kolb HC, Witte ON, Satyamurthy N, Heath JR, Phelps ME, Quake SR, Tseng H-R (2005) Multistep synthesis of a radiolabeled imaging probe using integrated microfluidics. Science 310(5755):1793–1796

Lee C-Y, Chang C-L, Wang Y-N, Fu L-M (2011) Microfluidic mixing: a review. Int J Mol Sci 12 (5):3263–3287

Liang SH, Collier TL, Rotstein BH, Lewis R, Steck M, Vasdev N (2013) Rapid microfluidic flow hydrogenation for reduction or deprotection of 18F-labeled compounds. Chem Commun 49 (78):8755–8757

Liu K, Lepin EJ, Wang MW, Guo F, Lin WY, Chen YC, Sirk SJ, Olma S, Phelps ME, Zhao XZ, Tseng HR, Michael van Dam R, Wu AM, Shen CK (2011) Microfluidic-based 18F-labeling of biomolecules for immuno-positron emission tomography. Mol Imaging 10(3):168–176, 161–167

Lu S-Y, Watts P, Chin FT, Hong J, Musachio JL, Briard E, Pike VW (2004) Syntheses of 11C- and 18F-labeled carboxylic esters within a hydrodynamically-driven micro-reactor. Lab Chip 4 (6):523–525

Lu S, Giamis AM, Pike VW (2009) Synthesis of [18F]Fallypride in a micro-reactor: rapid optimization and multiple-production in small doses for micro-PET studies. Curr Radiopharm 2(1):49–55

Lu S, Pike VW (2010) Synthesis of [18F]xenon difluoride as a radiolabeling reagent from [18F] fluoride ion in a micro-reactor and at production scale. J Fluor Chem 131(10):1032–1038

Ma X, Tseng W-Y, Eddings M, Keng PY, van Dam RM (2014) A microreactor with phase-change microvalves for batch chemical synthesis at high temperatures and pressures. Lab Chip 14 (2):280–285

Mae K, Maki T, Hasegawa I, Eto U, Mizutani Y, Honda N (2004) Development of a new micromixer based on split/recombination for mass production and its application to soap free emulsifier. Chem Eng J 101(1–3):31–38

Mansur EA, Ye M, Wang Y, Dai Y (2008) A state-of-the-art review of mixing in microfluidic mixers. Chin J Chem Eng 16(4):503–516

McMullen JP, Stone MT, Buchwald SL, Jensen KF (2010) An integrated microreactor system for self-optimization of a heck reaction: from micro- to mesoscale flow systems. Angew Chem Int Ed 49(39):7076–7080

Meille V (2006) Review on methods to deposit catalysts on structured surfaces. Appl Catal A Gen 315:1–17

Mengeaud V, Josserand J, Girault HH (2002) Mixing processes in a zigzag microchannel: finite element simulations and optical study. Anal Chem 74(16):4279–4286

Miller PW, Audrain H, Bender D, deMello AJ, Gee AD, Long NJ, Vilar R (2011) Rapid carbon-11 radiolabelling for PET using microfluidics. Chemistry 17(2):460–463

Miller PW, Long NJ, Vilar R, Gee AD (2008) Synthesis of 11C, 18F, 15O, and 13N radiolabels for positron emission tomography. Angew Chem 47(47):8998–9033

Mitchell MC, Spikmans V, Mello AJD (2001) Microchip-based synthesis and analysis: control of multicomponent reaction products and intermediates. Analyst 126(1):24–27

Moon B-U, Koster S, Wientjes KJC, Kwapiszewski RM, Schoonen AJM, Westerink BHC, Verpoorte E (2010) An enzymatic microreactor based on chaotic micromixing for enhanced amperometric detection in a continuous glucose monitoring application. Anal Chem 82 (16):6756–6763

Morini GL (2004) Single-phase convective heat transfer in microchannels: a review of experimental results. Int J Therm Sci 43(7):631–651

Murphy ER, Martinelli JR, Zaborenko N, Buchwald SL, Jensen KF (2007) Accelerating reactions with microreactors at elevated temperatures and pressures: profiling aminocarbonylation reactions. Angew Chem Int Ed 46(10):1734–1737

Nagaki A, Tomida Y, Yoshida JI (2008) Microflow-system-controlled anionic polymerization of styrenes. Macromolecules 41(17):6322–6330

Nguyen NT, Wu ZG (2005) Micromixers – a review. J Micromech Microeng 15(2):R1–R16

Ohkawa K, Nakamoto T, Izuka Y, Hirata Y, Inoue Y (2008) Flow and mixing characteristics of σ-type plate static mixer with splitting and inverse recombination. Chem Eng Res Des 86 (12):1447–1453

Okubo Y, Toma M, Ueda H, Maki T, Mae K (2004) Microchannel devices for the coalescence of dispersed droplets produced for use in rapid extraction processes. Chem Eng J 101(1–3):39–48

Panić S, Loebbecke S, Tuercke T, Antes J, Bošković D (2004) Experimental approaches to a better understanding of mixing performance of microfluidic devices. Chem Eng J 101(1–3):409–419

Pascali G, Mazzone G, Saccomanni G, Manera C, Salvadori PA (2010) Microfluidic approach for fast labeling optimization and dose-on-demand implementation. Nucl Med Biol 37(5):547–555

Pascali G, Nannavecchia G, Pitzianti S, Salvadori PA (2011) Dose-on-demand of diverse 18F-fluorocholine derivatives through a two-step microfluidic approach. Nucl Med Biol 38 (5):637–644

Pascali G, Watts P, Salvadori PA (2013) Microfluidics in radiopharmaceutical chemistry. Nucl Med Biol 40(6):776–787

Pascali G, Berton A, Rosaria DeSimone M, Wyatt N, Matesic L, Greguric I, Salvadori PA (2014) Hardware and software modifications on the Advion NanoTek microfluidic platform to extend flexibility for radiochemical synthesis. Appl Radiat Isot 84:40–47

Phelps ME (2000) Positron emission tomography provides molecular imaging of biological processes. Proc Natl Acad Sci 97(16):9226–9233

Razzaq T, Glasnov TN, Kappe CO (2009) Continuous-flow microreactor chemistry under high-temperature/pressure conditions. Eur J Org Chem 2009(9):1321–1325

Rebrov EV, Duisters T, Löb P, Meuldijk J, Hessel V (2012) Enhancement of the liquid-side mass transfer in a falling film catalytic microreactor by in-channel mixing structures. Ind Eng Chem Res 51(26):8719–8725

Rensch C, Jackson A, Lindner S, Salvamoser R, Samper V, Riese S, Bartenstein P, Wängler C, Wängler B (2013) Microfluidics: a groundbreaking technology for PET tracer production? Molecules 18(7):7930–7956

Richter S, Bouvet V, Wuest M, Bergmann R, Steinbach J, Pietzsch J, Neundorf I, Wuest F (2012) (18)F-Labeled phosphopeptide-cell-penetrating peptide dimers with enhanced cell uptake properties in human cancer cells. Nucl Med Biol 39(8):1202–1212

Riva E, Gagliardi S, Martinelli M, Passarella D, Vigo D, Rencurosi A (2010) Reaction of Grignard reagents with carbonyl compounds under continuous flow conditions. Tetrahedron 66 (17):3242–3247

Sadeghi S, Liang V, Cheung S, Woo S, Wu C, Ly J, Deng Y, Eddings M, van Dam RM (2013) Reusable electrochemical cell for rapid separation of [18F]fluoride from [18O]water for flow-through synthesis of 18F-labeled tracers. Appl Radiat Isot 75:85–94

Saito F, Nagashima Y, Goto A, Iwaki M, Takahashi N, Oka T, Inoue T, Hyodo T (2007) Electrochemical transfer of (18)F from (18)O water to aprotic polar solvent. Appl Radiat Isot 65(5):524–527

Schonfeld F, Hessel V, Hofmann C (2004) An optimised split-and-recombine micro-mixer with uniform "chaotic" mixing. Lab Chip 4(1):65–69

Schwalbe T, Autze V, Wille G (2002) Chemical synthesis in microreactors. CHIMIA Int J Chem 56(11):636–646

Schwarzer H-C, Peukert W (2004) Combined experimental/numerical study on the precipitation of nanoparticles. AIChE J 50(12):3234–3247

Schwarzer H-C, Schwertfirm F, Manhart M, Schmid H-J, Peukert W (2006) Predictive simulation of nanoparticle precipitation based on the population balance equation. Chem Eng Sci 61 (1):167–181

Serdons K, Verbruggen A, Bormans GM (2009) Developing new molecular imaging probes for PET. Methods 48(2):104–111

Singh BK, Kaval N, Tomar S, Eycken EVD, Parmar VS (2008) Transition metal-catalyzed carbon−carbon bond formation Suzuki, Heck, and Sonogashira reactions using microwave and microtechnology. Org Process Res Dev 12(3):468–474

Soleymani A, Kolehmainen E, Turunen I (2008) Numerical and experimental investigations of liquid mixing in T-type micromixers. Chem Eng J 135(Suppl):S219–S228

Sprogies T, Köhler JM, Groß GA (2008) Evaluation of static micromixers for flow-through extraction by emulsification. Chem Eng J 135(Suppl):S199–S202

Ståhl M, Åslund BL, Rasmuson ÅC (2001) Reaction crystallization kinetics of benzoic acid. AIChE J 47(7):1544–1560

Stroock AD, Dertinger SKW, Ajdari A, Mezić I, Stone HA, Whitesides GM (2002) Chaotic mixer for microchannels. Science 295(5555):647–651

Suga S, Nagaki A, Yoshida J-I (2003) Highly selective Friedel-Crafts monoalkylation using micromixing. Chem Commun 3:354–355

Suh YK, Kang S (2010) A review on mixing in microfluidics. Micromachines 1(3):82–111

Tarn MD, Pascali G, De Leonardis F, Watts P, Salvadori PA, Pamme N (2013) Purification of 2-[18F]fluoro-2-deoxy-D-glucose by on-chip solid-phase extraction. J Chromatogr A 1280:117–121

Thayer AM (2005) Harnessing microreactions. Chem Eng News 83(22):43–52

Ueno M, Hisamoto H, Kitamori T, Kobayashi S (2003) Phase-transfer alkylation reactions using microreactors. Chem Commun 8:936–937

van Herk D, Castaño P, Makkee M, Moulijn JA, Kreutzer MT (2009) Catalyst testing in a multiple-parallel, gas–liquid, powder-packed bed microreactor. Appl Catal A Gen 365 (2):199–206

Wang L, Liu D, Wang X, Han X (2012) Mixing enhancement of novel passive microfluidic mixers with cylindrical grooves. Chem Eng Sci 81:157–163

Watts P, Pascali G, Salvadori PA (2012) Positron emission tomography radiosynthesis in microreactors. J Flow Chem 2(2):37–42

Wheeler TD, Zeng D, Desai AV, Onal B, Reichert DE, Kenis PJ (2010) Microfluidic labeling of biomolecules with radiometals for use in nuclear medicine. Lab Chip 10(24):3387–3396

Wiles C, Watts P (2011) Recent advances in micro reaction technology. Chem Commun 47 (23):6512–6535

Wiles C, Watts P (2012) Continuous flow reactors: a perspective. Green Chem 14(1):38–54

Wilson NS, Sarko CR, Roth GP (2004) Development and applications of a practical continuous flow microwave cell. Org Process Res Dev 8(3):535–538

Wong R, Iwata R, Saiki H, Furumoto S, Ishikawa Y, Ozeki E (2012) Reactivity of electrochemically concentrated anhydrous [18F]fluoride for microfluidic radiosynthesis of 18F-labeled compounds. Appl Radiat Isot 70(1):193–199

Wu C-Y, Tsai R-T (2013) Fluid mixing via multidirectional vortices in converging-diverging meandering microchannels with semi-elliptical side walls. Chem Eng J 217:320–328

Zeng D, Desai AV, Ranganathan D, Wheeler TD, Kenis PJ, Reichert DE (2013) Microfluidic radiolabeling of biomolecules with PET radiometals. Nucl Med Biol 40(1):42–51

Ziegenbalg D, Löb P, Al-Rawashdeh MM, Kralisch D, Hessel V, Schönfeld F (2010) Use of "smart interfaces" to improve the liquid-sided mass transport in a falling film microreactor. Chem Eng Sci 65(11):3557–3566

Zuidhof NT, de Croon MHJM, Schouten JC, Tinge JT (2012) Beckmann rearrangement of cyclohexanone oxime to ϵ-caprolactam in a microreactor. Chem Eng Technol 35 (7):1257–1261

Index

J. Castillo-León, W.E. Svendsen (eds.), *Lab-on-a-Chip Devices and Micro-Total Analysis Systems*, DOI 10.1007/978-3-319-08687-3

MIX
Papier aus verantwortungsvollen Quellen
Paper from responsible sources
FSC® C105338

If you have any concerns about our products, you can contact us on
ProductSafety@springernature.com

In case Publisher is established outside the EU, the EU authorized representative is:
Springer Nature Customer Service Center GmbH
Europaplatz 3, 69115 Heidelberg, Germany

Printed by Libri Plureos GmbH
in Hamburg, Germany